$14.95

Ned DeLoach's Diving Guide to
UNDERWATER FLORIDA
SPRINGS • KEYS • PANHANDLE • EAST & WEST COASTS

W9-APL-487

Cover Photo: Paul Humann   **Design & Production:** Publication Arts, Inc.

Contributing Photographers:
Bob Burgess, Richard Collins, Ned DeLoach, Bill George, Danny Grizzard, Paul Humann, Tom Mount, Jim Ryan, Bill Smith, Roy Stoutamire, Steve Straatsma, Mark Ullmann, Pete Velde

Special Thanks:
Joanne Balch, Richard Collins, Mary DeLoach, Patricia Reilly-Collins, Joseph Gies, Michael O'Connell and Carol Yarborough

Published & Distributed by:
NEW WORLD PUBLICATIONS
1861 Cornell Road, Jacksonville, Florida 32207   Ph. (904) 737-6558
For an extra copy of *Underwater Florida* – mail $14.95 + 2.00 (Postage & Handling)

ISBN 1-878348-02-7

EDITOR'S NOTE

I came to Florida 21 years ago to dive. Why Florida? Simply because it was the place for divers to be and still is today. My first image of diving, the vision that sparked my lifelong search for underwater adventure, came from the state. It occurred during a family vacation in the mid-50s. My father was driving the coastal highway that skirts the Panhandle's beautiful beaches. Crossing a bridge over an inlet on the Gulf, I saw three barefoot boys, fins and mask in hand, carefully edging their way down a steep embankment toward the clear water below. That Christmas, I received my first set of flippers and a mask. In the summer I visited every drain in the bottom of every pool in my land-locked hometown in west Texas. Soon I began venturing into the dark waters of local rivers and lakes. After completing school I immediately packed up and headed east and have yet to be disappointed by what I have found. Florida's clear water, acres of living coral gardens, majestic wrecks and pristine spring pools all combine to offer a variety and quality of underwater wonders matched nowhere on earth.

This, the 7th edition of *Underwater Florida*, marks the 20 years since I began to publish diving guides to the state. The first, a small 48-page booklet entitled *Diving and Recreational Guide to Florida's Springs*, appeared at the end of 1971. It was born from my frustrating efforts to locate spring sites to explore. Two years later, I spent the summer camping and diving in the Keys to gather material for a second guide, *Diving Guide to the Florida Keys*. In 1977 the two books were combined and sites from the east and west coast were added to create the present format of *Underwater Florida*. This current edition includes 30 chapters that give directions and information for over 500 of Florida's best dives.

The effort of compiling this guide was not mine alone. Many charter boat captains, dive store operators, and agencies that direct the formation of artificial reef sites have spent countless hours graciously sharing experiences, observations and information so that their personal love and excitement for Florida's underwater realm could be passed on to others.

A fear that has disturbed each of us has been that our natural wonders would be overused and abused to such an extent that they would be lost to future generations. We well know that the battle to save our seas is never ending, but it has become evident what diligent effort by educated, caring, understanding men and women can accomplish. Today, our guarded optimism burns brighter than it did two decades ago.

The very best have come forward to rationally and persistently act as guardians of our waters. Through their efforts and those of county, state and federal agencies, much has already been preserved. The two marine sanctuaries in the Keys have shown the success possible by proper management of our marine resources. Hopefully, they will be the first of many such areas in our state and federal waters. Costly artificial reef development programs from Pensacola to Miami have created abundant sea life where none existed before. Organizations for the protection of our reefs have established systems of mooring buoys, greatly limiting damage from boat anchors. Dedicated groups protecting sea turtles and manatees have won the hearts and concerns of our citizenry. The moratorium on offshore oil exploration was a hard won battle by dedicated individuals and conservation groups. These groups represent the future of diving.

However, the realization of our hopes rests with the individual divers and watersportsmen who, through their ever-growing awareness and efforts, are becoming the sea's most effectual ambassadors.

This book is dedicated to those who, by care and understanding, have already achieved so much, yet realize that our vigil is one that is never ending.

HOW TO USE THIS GUIDE

Underwater Florida is designed to give divers a complete overview of the many diving opportunities found in Florida waters. There are 30 chapters; the area included in each chapter is shown on the Florida Map, pages 8 & 9. The Map Index, page 8, lists the page numbers where the chapter information begins.

From a list of many springs dived, I have included only those sites that are currently accessible to the public and provide the most rewarding diving experiences. Some are commercial, state or county operated diving parks that charge entrance fees. Nearly all the springs listed can be reached by car; however, many are difficult to find because of their remote locations. Directions are given by highways and roads. (Many are nothing more than roughly formed sand paths and a few become impassable during periods of heavy rainfall.) The distances in miles and tenths of a mile, which can be read on an automobile odometer, are usually given from main highway intersections or bridges.

Directions and information are also included for saltwater diving sites that do not require a boat to reach. These areas are marked "Beach Dive" after the location's name. A special listing of beach dives is included in the index. Rewarding diving without a boat is found at places in the Panhandle, the state's southwest coast and along the mid-to-lower east coast. Beach diving in the Keys is virtually non-existent. Reef areas are too far offshore, and strong current flows between islands make bridge diving hazardous.

The maps and directions for offshore locations are only general references. The Loran-C numbers, although of great help in offshore navigation, vary in accuracy from unit to unit. They nearly always need to be used in connection with other navigational aids, as well as a depth finder to pinpoint the exact location you are searching for.

Most divers generally prefer to use charter diving services when going offshore. Many such services are advertised in this book.

I hope that the information found in this diving guide will save you time and effort in your search for the beauty and adventure found in Florida's underwater world.

Hal Watts,
(MR. SCUBA)
Founder of
The Professional Scuba Association (PSA)
invites you to try a Unique Diving
Experience at **FORTY FATHOM GROTTO**
Ocala, FL.

Professional Scuba Association psa Inc
2219 E. Colonial Dr., Orlando, FL 32803 • (407) 896-4541

Highway Mileage Chart for Popular Dive Locations and Major Cities in Florida

	Alexander Springs	Blue Grotto	Blue Spring—Orange City	Branford	Crystal River	Forty Fathom Grotto	Ft. Myers	Gainesville	Ginnie Spring	Ichetucknee Spring	Jacksonville	Key Largo	Little River Spring	Merritt's Mill Pond	Miami	Morrison Spring	Orlando	Paradise Spring	Peacock Springs	Panama City	Pensacola	River Rendezvous	Tallahassee	Tampa	Troy Spring	Vortex Spring	West Palm
Alexander Springs	—	66	35	127	90	40	203	77	97	105	98	320	132	288	282	329	50	38	150	305	406	140	213	111	134	334	216
Blue Grotto	66	—	89	61	45	17	232	23	42	54	102	366	66	208	328	230	102	37	84	237	338	74	141	100	68	246	266
Blue Spring—Orange City	35	89	—	145	113	75	183	100	131	158	106	300	150	311	262	352	30	72	168	328	429	158	236	115	152	357	196
Branford	127	61	145	—	87	86	234	47	35	8	84	420	161	311	382	200	156	106	23	204	305	13	94	110	7	201	317
Crystal River	90	45	113	87	—	55	208	63	83	88	133	360	92	233	220	272	90	45	110	259	305	99	166	63	90	273	254
Fourty Fathom Grotto	40	17	75	86	55	—	208	45	70	80	99	354	91	200	316	239	84	51	109	245	346	48	133	102	93	238	250
Ft. Myers	203	232	183	234	208	208	—	231	263	275	285	173	239	423	145	200	153	191	257	448	549	193	356	124	238	463	124
Gainesville	77	23	100	47	63	45	231	—	32	44	70	373	52	211	335	252	109	65	70	236	337	112	144	128	54	252	270
Ginnie Spring	97	42	131	35	83	70	263	32	—	27	88	405	40	196	367	235	124	90	58	239	340	47	129	139	42	236	285
Ichetucknee Spring	105	54	158	8	88	80	275	44	27	—	80	432	13	169	379	208	134	107	31	212	313	20	102	151	15	209	309
Jacksonville	98	102	106	84	133	99	285	70	88	80	—	398	89	229	345	269	134	106	95	261	360	94	168	190	91	270	277
Key Largo	320	366	300	420	360	354	173	373	405	432	398	—	425	472	62	614	270	334	443	598	699	432	506	287	427	613	122
Little River Spring	132	66	150	161	92	91	239	52	40	13	89	425	—	166	387	125	161	111	15	209	310	17	99	115	12	209	322
Merritt's Mill Pond	288	208	311	311	233	200	423	211	196	169	229	472	166	—	534	40	309	220	154	54	131	153	67	306	156	49	469
Miami	282	328	262	382	220	316	145	335	367	379	345	62	387	534	—	575	232	296	405	560	661	394	467	249	389	575	68
Morrison Spring	329	230	352	200	272	239	200	252	235	208	269	614	125	40	575	—	348	259	193	52	89	212	106	346	195	10	508
Orlando	50	102	30	156	90	84	153	109	124	134	134	270	161	309	232	348	—	64	179	334	435	168	242	85	163	347	166
Paradise Spring	38	37	72	106	45	51	191	65	90	107	106	334	111	220	296	259	64	—	129	265	366	153	85	85	113	258	230
Peacock Springs	150	84	168	23	110	109	257	70	58	31	95	443	15	154	405	193	179	129	—	207	308	6	87	130	19	192	340
Panama City	305	237	328	204	259	245	448	236	239	212	261	598	209	54	560	52	334	265	207	—	101	195	97	332	199	61	494
Pensacola	406	338	429	305	360	346	549	337	340	313	360	699	310	131	661	89	435	366	308	101	—	295	198	433	300	91	595
River Rendezvous	140	74	158	13	99	98	193	112	47	20	94	432	17	153	394	212	168	153	6	195	295	—	85	118	101	190	327
Tallahassee	213	141	236	94	166	133	356	144	129	102	168	506	99	67	467	106	242	85	87	97	198	85	—	240	101	105	402
Tampa	111	100	115	110	63	102	124	128	139	151	190	287	115	306	249	346	85	85	130	332	433	118	240	—	112	347	195
Troy Spring	134	68	152	7	90	93	238	54	42	15	91	427	12	156	389	195	163	113	19	199	300	101	101	112	—	197	324
Vortex Spring	334	246	357	201	273	238	463	252	236	209	270	613	209	49	575	10	347	258	192	61	91	190	105	347	197	—	509
West Palm	216	266	196	317	254	250	124	270	285	309	277	122	322	469	68	508	166	230	340	494	595	327	402	195	324	509	—

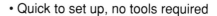

CONTENTS

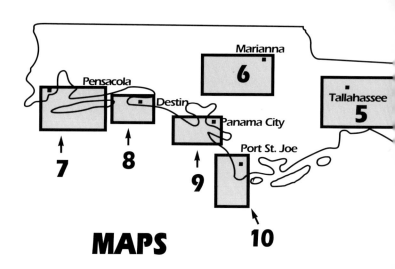

MAPS

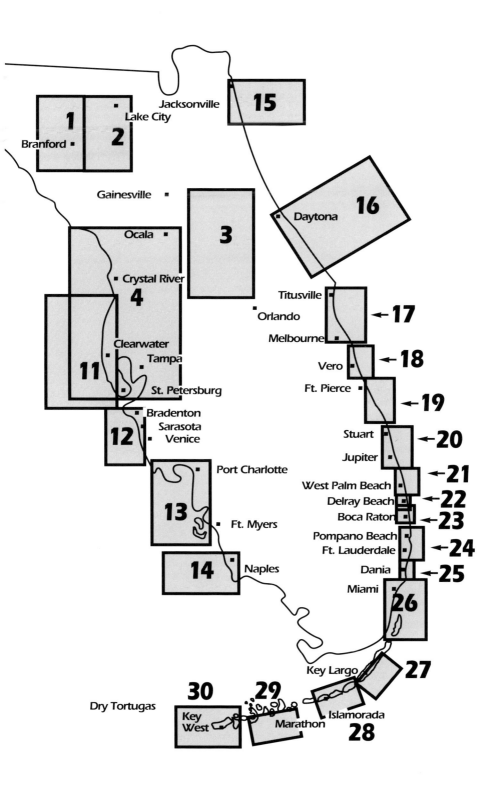

1

Branford •

2

Jacksonville

Lake City

15

Gainesville •

Ocala •

3

Daytona

16

• Crystal River

4

11

Clearwater

Tampa

St. Petersburg

Titusville

• Orlando

17

Melbourne

18

Vero

19

Ft. Pierce •

Bradenton

Sarasota

Venice

12

Port Charlotte

13

Ft. Myers •

14

Naples

Stuart

20

Jupiter

21

West Palm Beach

22

Delray Beach

23

Boca Raton

Pompano Beach

Ft. Lauderdale

24

Dania

25

Miami

26

Key Largo

27

30

29

Dry Tortugas

Key
West

Marathon

Islamorada

28

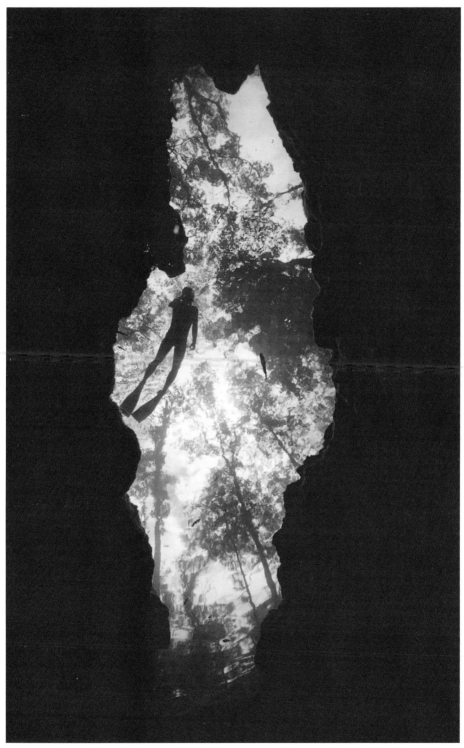

Little Devil at Ginnie Springs.

Ned DeLoach

UNDERWATER ADVENTURE BY AUTOMOBILE

Four Dream Tours Through Florida's Underwater Wonderlands

Florida has it all—reefs, wrecks, springs, caves, rigs for diving, fish watching, lobstering, relic hunting, shell collecting, drift diving, underwater photography, treasure hunting—everything but ice diving.

In the ocean waters that skirt Florida's 1,300 miles of coastline and in her unbelievably clear spring pools, scenic rivers and placid lakes, underwater explorers will discover more spectacular and varied marine environments than any other place on earth. This fact, coupled with the ability to drive rather than fly to the state's hundreds of popular underwater sites, has long made Florida the world's most popular diving destination.

The use of an automobile provides the freedom necessary to search out a diversity of superb marine terrains in just a matter of days. With the correct directions you can drive to within a few yards from dozens of spring basins or river sites and right up to an intracoastal marina where one of the state's fleet of nearly 200 diving boat charter services awaits.

Three great interstate roadways bring divers from across America to the Florida peninsula—from the west I-10, from the heartland I-75, and paralleling the eastern seaboard south, I-95. When you enter the state on either highway, you still have nearly 400 miles to travel before you arrive at North America's stellar underwater attraction—the living coral reefs of South Florida and the Keys. Although the coral gardens are not to be missed, many other premier diving sites are located along the paths south. Too many visitors are unaware of the diving potential spread throughout the state and miss much of the underwater diversity in Florida waters.

To help the traveling diver get a proper perspective of the exciting driving vacations that can be made throughout the peninsula, I have planned three dream itineraries. Each begins with the divers entering the state on one of the three interstate highways. All end in the West Palm area on the lower east coast. From there the journeys will take a common route south through Miami and along the entire 108-mile length of the Florida Keys ending at Key West.

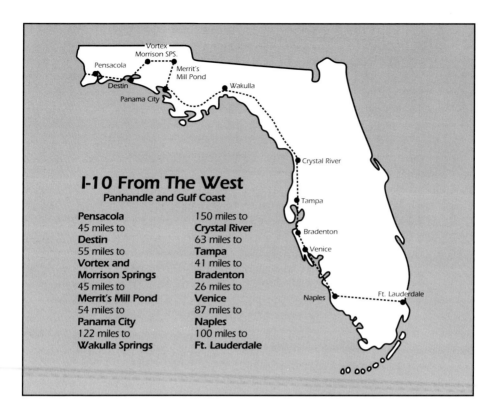

I-10 From The West
Panhandle and Gulf Coast

Pensacola
45 miles to
Destin
55 miles to
**Vortex and
Morrison Springs**
45 miles to
Merrit's Mill Pond
54 miles to
Panama City
122 miles to
Wakulla Springs

150 miles to
Crystal River
63 miles to
Tampa
41 miles to
Bradenton
26 miles to
Venice
87 miles to
Naples
100 miles to
Ft. Lauderdale

The Panhandle and Gulf Coast

From the west, I-10 enters Florida's panhandle just north of Pensacola. This 200-mile neck of land from the western border to the state's capital at Tallahassee is a true sportsman's paradise. Still under development, the region is a wild tangle of pine and hardwood forests and lowlands that are crisscrossed by dozens of cypress-lined rivers and spring-fed creeks. To the south, the Gulf of Mexico rims the land with many of America's most beautiful beaches. A sea drive west from Pensacola through Destin and Panama City will never be forgotten. Wide white beaches, buttressed by miles of rolling dunes, separate forest from sea.

Unfortunately, beach diving is disappointing along the coast. Shallow sand flats stretch monotonously for miles, with only occasional sand dollars or starfish to be found. The real underwater adventure starts a few miles out to sea. There, great shipwrecks grace the ocean's floor. Through the years these towering steel spires have enticed myriad marine life from the open water. The summer and early fall months are best for diving in the Gulf. Seas are generally calm at this time and visibility hovers around 60 feet.

Excellent dive boat charter services are located in Pensacola, Destin and Panama City, and Port St. Joe. Many people take their first dive out from Pensacola. The USS *Massachusetts* is an interesting shallow water dive. The 500-foot battleship has been down since 1927. She is only a mile out from the rock jetties in 25 feet of water. Other area sites include a 480-foot Liberty Ship in 80 feet and the popular Russian freighter *San Pablo* that was torpedoed in the Florida Straits during World War II.

Just 40 miles east on US 98 is the lovely community of Destin, a favorite beachfront playground for tourists during the spring and summer. Some of the Gulf's most spectacular rock reefs are found just offshore. These high limestone walls, remnants of forgotten shorelines, run for hundreds of feet. Most are undercut by deep crevices and small caves where game fish lurk. Besides the great ledges, there are many exciting wrecks to explore.

After diving at Destin you can experience a dramatic change of underwater scenery by heading northwest up US 331 for 55 miles. Here, deep within the woodlands, three beautiful basins await. Morrison, Cypress and Vortex Springs are commercially operated diving parks located less than ten miles apart. Such spring pools offer the diver the clearest water imaginable.

Next, follow I-10 west for 45 miles. Just outside Marianna is Merrit's Mill Pond, a spring-fed lake five miles long with half a dozen spring areas to discover. The lake waters flow two miles down Spring Creek to the Chipola River, one of the best underwater artifact hunting areas in the state. From here head south once again to Panama City for some more Gulf diving. Don't miss the thrill of visiting such legendary wrecks as the *Tarpon*, or *Grey Ghost*. Adding even more variety and excitement to the diving are the recent additions of the 205-foot tug Chippewa and the eighteen bridge span section of the old Hathaway Bridge. After Panama City stay on US 98 as it hugs the coast southwest, to the small fishing village of Port St. Joe. In the Gulf waters just out from Cape San Blas are a dozen good dives, including a Florida diving legend, the *Empire Mica*. The 480-foot British tanker was torpedoed by a U-boat in 1942. Today her broken remains rest in 100 feet of clear water.

From here the coastal drive passes the renowned oyster grounds of Apalachicola Bay. There is no diving along this stretch, but the memories from this ocean drive will be lasting. After 120 miles, make a two-mile detour to Wakulla Springs. This picturesque wildlife refuge, purchased by the state in 1986 is the home one of the world's largest waterfilled caves. No scuba is allowed here, much to the chagrin of Florida cave divers who for decades have coveted a dive into the spectacular cavern.

It is less than 150 miles south to the small but growing town of Crystal River on the banks of Kings Bay. The waters here are the winter home of large herds of manatee. For centuries, when the ocean's temperatures drop, these loveable, lumbering mammals gather in the bay's clear 72-degree water. The pleasant water comes from over 30 spring vents. Now divers far outnumber the manatees. Though always crowded on weekends, Crystal River is your best bet to swim with one of these gentle sea cows that everyone adores.

West coast ocean diving starts 45 miles south at Hudson and continues through the Tampa area and on to Sarasota. These waters don't have tropical coral gardens or crystalline seas, but they do have things that send the spirits of underwater explorers racing: game fish—large, bold and abundant—extensive, well-maintained artificial reefs, great wrecks, and always great adventure.

Nearly all diving must be done from a boat. However, there are a few good spots where beach dives can be made. Off the Bradenton Beach, the Sugar Barge and Third Pier are areas that have been dived for years. South, off Boca Grande Island, there are several more good spots. Another popular beach dive can be made out from Venice Public Beach. Here divers search for giant prehistoric shark's teeth.

Thousands of specimens, up to six inches across, have been found by the patient hunter. What a great souvenir they make for a vacationing diver!

From the southwest coast at Naples, take the 100-mile run across Alligator Alley, the thoroughfare that cuts its way east across the waving grass sea called the Everglades. This will take you to Ft. Lauderdale on the lower east coast, in the heart of South Florida's renowned clearwater diving mecca.

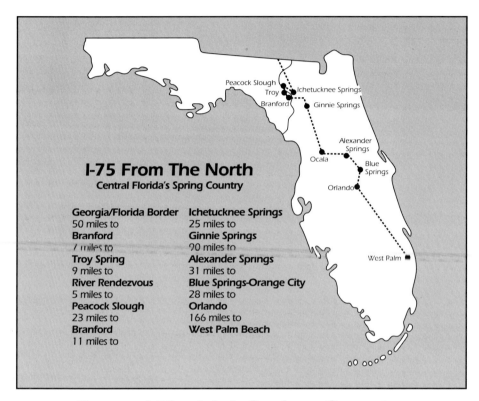

I-75 From The North
Central Florida's Spring Country

Georgia/Florida Border	Ichetucknee Springs
50 miles to	25 miles to
Branford	**Ginnie Springs**
7 miles to	90 miles to
Troy Spring	**Alexander Springs**
9 miles to	31 miles to
River Rendezvous	**Blue Springs-Orange City**
5 miles to	28 miles to
Peacock Slough	**Orlando**
23 miles to	166 miles to
Branford	**West Palm Beach**
11 miles to	

Central Florida's Spring Country

Fifteen miles after I-75 enters north Florida from Georgia take the US 129 exit. This route will carry you almost directly south for 35 miles to Branford on the Suwannee River's eastern bank. For 30 years this small, unassuming farming community has been the staging center for diving expeditions into Florida's famous spring diving country. Within a 20-mile radius are more than 30 diveable springs and sinkholes of which eight must rank as the best in the world. For the most part these spring pools are situated in undeveloped woodland along the banks of the Suwannee and nearby Santa Fe Rivers. All can be reached easily by car.

Although these placid, clear pools offer the safest diving conditions imaginable, they have gained an undeserved reputation for danger because of the drownings that occur in their caves that channel the ground water up into the basins. At many spring sites these openings are large enough for divers to enter easily. When those divers who lack the advanced training required to safely circumnavigate their intertwining passages penetrate the system beyond the glow of natural light, they immediately enter a hazardous situation. Few realize the danger imposed by a cave's ceiling. If you plan to dive the caves, be sure to receive the specialized cave training first from one of the areas dive operator.

Days can be spent in the Branford area diving, canoeing, fishing and camping. It would be impractical to attempt a dive at all the springs during a single visit, but in three days you can cover most of the hot spots.

On the first day, head for Troy Springs, which is located seven miles northwest of Branford. This large, open basin is within a stone's throw from the Suwannee's west bank. Steep walls lead down to water's edge. They continue their plunge underwater for more than 80 feet. At the bottom, several small vents issue millions of gallons of cool water that constantly refresh the basin. Down the short run near the river's dark waters are the broken ribs of a 19th century steamboat that was scuttled here during the Civil War.

A few miles up the highway is River Rendezvous, an inland diving park. This lodge, restaurant, campground, dive shop complex was built on a high bluff overlooking the river at one of its most scenic points. In front of the lodge is a spring pool about 18 feet deep.

After crossing the river on the S-51 bridge, take a right at the second road. This will lead to a new state park that includes Peacock Slough—a series of seven beautiful springs and sinkholes connected by an extensive network of underground passages. Like all state facilities that offer diving, certification is required, and only those with advanced cave training are permitted to carry lights into the caverns. This rule forces untrained cavern divers to stay within sight of daylight. The best dives here are Peacock Spring and Orange Grove Sink located just a mile away. From the Peacock area to Branford there are several more spring sites on the river's western bank. You have Cow, Running, Royal or Little River Spring to choose from. Each is an

Diver enters the small spring-siphon, Cow Spring.   Ned DeLoach

Canoeing the Ichetucknee, a delight of Florida's spring country.   Ned DeLoach

individual delight.

The next day go 12 miles east on US 27 to the Ichetucknee State Park. From the headspring, three miles above the highway, you can spend a lazy afternoon drifting with a 2-knot current through a woodland paradise. Most visitors choose tire tubes for the float downstream. Divers enjoy snorkeling the crystalline run searching the bottom for relics and observing the abundant freshwater marine life. A wetsuit is recommended for the two to three hour trip as protection against the cool 72-degree temperature.

The following day go a few miles west to the Santa Fe River springs. The best dive here, without a doubt, is Ginnie Springs. The spring is located in a commercially operated diving/camping park on the river's bank. Several other springs, including Devil's Eye, are on the grounds. Although the entrance fee might seem a little steep after you have dived free for the past few days, you will find that the superb diving and excellent facilities are well worth the price. Ginnie is not to be missed.

After leaving Ginnie, travel a few miles west to get back on I-75 and drive 50 miles south to Ocala. From here take SR 50 west to the Ocala National Forest. This highway will carry you past the tourist attractions at Silver Springs. Diving and swimming are not allowed, but there is much to see, including an excellent petting park for the children. In the forest there are four spring areas worth a visit. Juniper, Salt, Silver Glen and Alexander all allow snorkeling. Scuba is permitted at Alexander and at Blue Springs just west of the forest near DeLand. Canoeing is excellent at all the spots.

You're now just north of Orlando and nearby Disney World. From there you can catch the Florida Turnpike and in less than three hours arrive at the West Palm area where underwater recreation is a way of life.

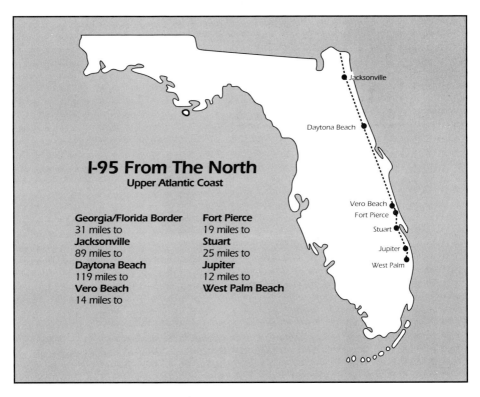

I-95 From The North
Upper Atlantic Coast

Georgia/Florida Border	**Fort Pierce**
31 miles to	19 miles to
Jacksonville	**Stuart**
89 miles to	25 miles to
Daytona Beach	**Jupiter**
119 miles to	12 miles to
Vero Beach	**West Palm Beach**
14 miles to	

Upper Atlantic Coast

I-95 enters northeast Florida just above Jacksonville and runs down the peninsula's east coast for 400 miles to Miami. The coastal zone for the first 200 miles has never been recognized as suitable for sport diving. For decades the waters were rumored to be deep, dark, rife with currents and shark infested. The story went that the ocean's floor spreads unbroken for miles, offering no protection for sea life and little to see. Few were willing to brave the long boat rides and uncertain underwater environment for a day's diving. However, with the advent of Loran C to help boat captains quickly pinpoint the exact locations of rock ledges and artificial reefs and the recent success of two sport diving charter services out of Jacksonville, things have changed. A new dive destination has once again been discovered in Florida waters.

The upper east coast offers a different type diving and it is not for everyone, but those who choose to explore the seas from Jacksonville through Daytona Beach and on to Melbourne can plan on being thrilled. These undiscovered waters are boiling with marine life in numbers and size that is rarely seen anywhere.

The term "wild diving" could aptly apply to underwater activities in this area, not because of danger but because of the unpredictability of what might be encountered when divers drop below the surface. On their way to the bottom they first pass through the open ranges of pelagics. This is where the action is, where the big, sleek, silver fish swim. When these fish are on the move, when the cobia, kings and tarpon swim free, there is not a more exciting place for a diver to be. It is a rare day that visitors are not investigated by inquisitive barracudas or stampeded by fast-swimming amberjacks while dropping down the anchor rope through this mid-water zone. Dozens of 30 to 40 pound fish will, without warning, sail into visibility heading directly toward the diver. Suddenly, at the last moment, they alter their course silently, flying past only feet away on every side.

Nearing the bottom, the profile of a limestone ledge takes form in the deep water haze. This is the diver's destination, the faint blip on the sounding screen that he has come to investigate. Such rare breaks, in an otherwise flat sea of sand, generally rise less than four feet off the bottom, but they churn with life. Everything from tiny multi-colored tropicals to large, bold game fish flock to hundreds of these marine oases. At least two dozen fish species can be quickly identified on every dive.

For years derelict seagoing vessels have been added to the flat seabed to attract sea life. These artificial reefs have also become popular dive sites. The first thing a diver will usually encounter when approaching a wreck location is a spiraling vortex of baitfish that moves as a cloud over the area. Schooling spadefish, each as large as a dinner plate, constantly maraud these legions. Divers should be alert when first approaching the tangled wreckage. This will be their chance to spot the big boys—jewfish to 300 pounds. Inside the sea-encrusted wreckage, thousands of life forms have taken refuge. Every crevice and cranny has become something's home. Here you will see sights you have never imagined.

These are the reasons why divers travel ten miles from shore and drop 90 feet into uncertain waters. It is a spectacular marine environment seen by few but never forgotten by those who venture there.

Farther down the east coast, beginning just north of Vero Beach and continuing to Jupiter, is one of the state's most popular beach diving area. Ocean access is convenient from any of a series of well-maintained public parks that front the Atlantic. From these entry points, divers can swim the 75 to 150 yards to a section of rock reef that parallels the shoreline. Depths vary from six to 15 feet. A variety of soft corals and sponges decorate the outcroppings while tropicals by the dozens dance over the ledges. Although these sites are not as dramatic as the reefs, farther offshore they do provide divers an excellent opportunity to visit a marine environment without the expense and scheduling problems associated with boat charters. The most popular dives near Vero are Wabasso Beach, Indian River Shores, the *Breconshire* Wreck and Round Island. North of Ft. Pierce, Pepper Park, Inlet Park and the Paddle-Wheeler Wreck are favorites. Around Jupiter, Cove Park and Carlin Park are recommended.

Beach diving is best here when the wind comes from the northeast. The surf can at times be calm as a lake. With less surge, water visibility improves markedly. Be sure to check the tide tables so that you will not have to return to the beach during an outgoing tide. Always trail a float (such as an inflated tire tube) with a large dive flag to mark your underwater location. Boat traffic can be expected on all sections of inshore reefs.

This section of the Atlantic Coast is also famous for huge "bull lobster." Many of the largest spiny lobster ever taken came from this area. Be on the lookout for these critters on every dive. Many have been caught that weigh over 13 pounds. If you want some real fun, join in on the annual lobster hunts sponsored by local dive shops. It is not only a chance to cover your grill with tasty tails, but also to win one of the dozens of prizes offered.

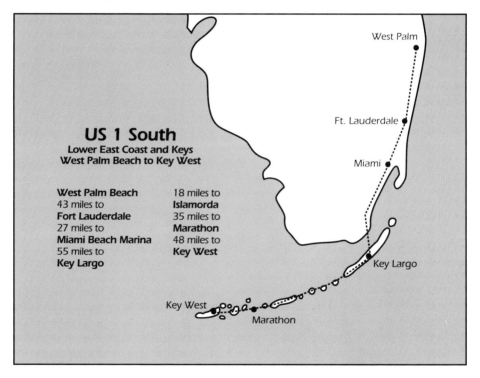

US 1 South
Lower East Coast and Keys
West Palm Beach to Key West

West Palm Beach	18 miles to
43 miles to	**Islamorda**
Fort Lauderdale	35 miles to
27 miles to	**Marathon**
Miami Beach Marina	48 miles to
55 miles to	**Key West**
Key Largo	

The Lower East Coast and Keys

South Florida's fabled clearwater diving begins off Palm Beach. Here the warm currents from the Gulf Stream course close to shore. The result is a year-round underwater wonderland for divers. A constant flow of clear, blue water continually feeds and refreshes the reefs, keeping them well-stocked with lobster, game fish, sea fans and small stands of hard coral. These currents enable divers to effortlessly drift over vast sections of the great reefs on single dives. On such marine odysseys it's common to spot sea creatures by the thousands. Another convenience of Palm Beach diving is that all reefs, ranging in depth from 30 to 100 feet, are within two miles of land. Several magnificent wreck dives can also be made here. The most famous is the 185-foot Greek luxury liner *Mizpah*. She's completely intact, resting in 90 feet of water. During the over 20 years the ship has been down, sea life has flocked to her for sanctuary. She's a sight to see!

The water south is also filled with memorable dives. An exceptional beach dive can be made on the Delray Wreck just off the south end of Delray's public beach. The steel-hulled freighter that sank in the 1920's has become popular with advanced as well as beginning divers. Her remains are in three distinct sections in 22 feet of water.

Another good shore entry dive is from the one-mile stretch of beach at Lauderdale-by-the-Sea. The best spots are 100 yards both north and south from the pier's end in 30 feet of water. The reef rises 4 to 8 feet off the sea bed and is adorned with waving sea fans and soft corals. This is a good place to night dive from the beach. You can mark your position by the lights on the pier and from hotels along the beach. It would be a good idea to attach warning strobe lights to your required surface float.

Although local divers in Ft. Lauderdale have known for decades about the good diving off the city's coast, their reefs remained undiscovered by vacationing divers until a bizarre happening occurred on a stormswept night in 1984. That evening, the Mercedes, a 108 foot froighter, was caught in galeforce winds and driven onto a beach near the pool of a prominent Palm Beach socialite. The event was well-recorded by the national TV networks. After a three month saga the derelict ship was finally freed. She was moved down the coast and put to rest off Ft. Lauderdale as an artificial reef. Overnight, the Lauderdale waters became a popular diving destination. The once secret reefs became common knowledge. One of the most beautiful sets of reef on the entire east coast was discovered. Ft. Lauderdale is no longer being passed up by southbound divers.

Miami diving is wreck diving at its best. Only a few miles off the city's shore, where the shallow bottom suddenly angles to great depths, rests the wreckage of a fleet of vessels, now artificial reefs. Each was carefully scuttled so that it would rest on the edge of the Gulf Stream to attract marine life. Clear, rich water constantly feeds the steel reefs. In only a few years the raw metal has become enveloped in a luxuriant carpet of coral and sponge that swarms with fish. Because such clear water bathes the wrecks, their visual impact is startling. Visibility does vary, depending on the Gulf Stream's caprice, but it often extends over 100 feet. All the wrecks lie relatively deep—from 70 to 200 feet. Those wishing to visit should be experienced open-water divers. A few of Miami's wrecks with their size and depth are: *Orion*, 118-foot tug in 65 feet; *Blue Fire*, 175-foot freighter in 80 feet; *Almirante*, 210-foot steel freighter in 100 feet. Each is a never-forgotten sea sculpture shaped by man, fashioned by nature.

US 1 doesn't stop at continent's end; instead, it continues for over 100 miles to sea, linking 31 islands with 40 bridges. These are the famous Florida Keys—an island world that you can drive to. The chain runs west, separating the shallow flats of the Gulf of Mexico from the Florida Reef that lies on the Gulf Stream's fringe in the Atlantic. This reef, the only living coral garden in North America, draws divers by the thousands to these waters.

After crossing Jewfish Creek, you arrive in Key Largo, the home of the Key Largo National Marine Sanctuary and Pennekamp Coral Reef State Park, where over 200 miles of underwater splendor has been protected since 1960. The result is a diver's paradise.

All good diving is well offshore. A boat is required. A dozen full-service diving charter businesses operate in the park. The captains know the waters and what the diver wants. Most take half-day excursions that leave in the morning and early afternoon. Generally, two dives are made in depths from 20 to 40 feet. Three of my favorite spots in the park are Carysfort Reef, located near the southeast boundary and marked by a 100-foot tower, with a shallow elkhorn garden and a beautiful coral-encrusted slope that drops to 65 feet; French Reef, and nearby *Benwood* wreck, both dramatic dives.

Pennekamp is only the beginning of the underwater adventure that awaits in the Keys. The sunken remains of some of the richest treasure galleons ever discovered can be found off the Upper Keys. In the past years, their high piles of egg-shaped ballast stones have given up thousands of silver coins and artifacts to hardworking treasure hunters. Much still remains hidden just under the acres of shifting sand. Adding to the area's romance is a collection of modern-day wrecks that grace the sea floor just off Islamorada. In 1985 the 287-foot freighter *Eagle* was sunk. This is a dive that shouldn't be missed.

The city of Marathon is the home of diving in the Middle Keys. Miles of thriving reef, 16th century galleon sites and a great wreck, the *Thunderbolt*, offer enough to keep divers happy for weeks. Just west is the Looe Key National Marine Sanctuary. This exciting reef area is possibly the most beautiful in the entire island chain. The V-shaped shoal is completely awash and varies in depth from two to 40 feet.

For years, divers were waylaid by the beautiful reefs of the Upper and Middle Keys and just did not seem to make it to Key West. To me the world just wouldn't be complete if there wasn't a Key West. It is well worth the drive. Diving here is also special. Starting at Western Sambo and continuing 45 miles westward to the Marquesas is a series of reefs covered with an abundance of life. Key West is also a point of departure for the Dry Tortugas, the Marquesas, and other out islands in the Gulf of Mexico, where underwater discoveries are made every day.

The diving sites mentioned here are only a sampling of over 500 locations described in Underwater Florida. There are dozens of other good dives located along each route that are worth your time to investigate. Read through the book carefully to become familiar with all the state's underwater wonders that await the traveling diver.

Florida—the affordable dive destination where American divers live their dreams.

Alex Kuze with gold plate he found while lobster diving   Jim Ryan

FLORIDA TREASURE HUNTING

THAR'S GOLD IN THEM THAR BEACHES

BY BOB BURGESS

Understandably, nobody talks much about it, and some people living there swear they never heard about it. But the truth is that certain southeast coast Florida beaches are producing gold and silver...and more of it is being found than you would imagine.

It's Spanish treasure, sometimes an emerald-studded gold cross, or a priceless religious relic; other times it may be a scattering of gold coins, or a corroded mass of silver. More often than not the beaches produce silver cobs—crudely clipped and stamped silver coins blackened with silver sulphide corrosion camouflaging their identities. Often mistaken by beachcombers for sun-baked chunks of tar, or corroded scrap metal, many are skipped back into the ocean. Talk about throwing away money! Cobs are worth anywhere from $75 and up apiece. Most of these are dated prior to 1715. In some areas, larger silver coins, bearing the first milled edges and called Pillar Dollars, have turned up. Dated 1732, these rare Spanish coins command prices of over $2,000. The Spanish gold coins are worth several thousands and up. Then come the exotics, the elaborate gold jewelry with the sky the limit. It is there, scattered, and under the right conditions, divers with proper equipment who narrow their searches to the right beaches, at the right time, are finding it.

Escudo gold coins valued at over $6,000 each   Bob Burgess

WHERE DOES IT COME FROM? Florida's coasts are littered with the remains of old shipwrecks. Thanks to the horrendous storms that periodically swooped through the Florida Straits over centuries of sailing history, occasionally destroying entire fleets of sailing ships, this breezeway with its tempestuous counter-currents has spelled doom for thousands of vessels. Two Spanish treasure fleets that perished there are responsible for today's major treasure finds, both offshore and on the beaches. The best beach sites for coinshooting treasure hunters are directly opposite the shipwrecks of the 1715 fleet that perished between Sebastian Inlet and just south of Fort Pierce. In that distance of some 20 miles were scattered over 10,000,000 pesos of Spanish treasure. Several of the ships grounded close ashore and their cargoes scattered far and wide along miles of ocean bottom as the storm surf and soft slurry of coastal sands gobbled up all traces of the tragedy.

By the mid-1960s, building contractor Kip Wagner had found enough early 18th century Spanish coins on the beach between Wabasso and Sebastian Inlet to convince him that a valuable Spanish treasure wreck lay offshore. With the help of scuba diving friends, that wreck and others scattered to the south were found and salvaged under contract with the State of Florida. The rewards were enormous. Literally tons of silver coins came from the northernmost wreck, called the Cabin Wreck, near Sebastian Inlet. These wreck sites are still being worked under special contracts. And according to prevailing laws, everything in the water, from the high tide mark out to Florida's 3-mile limit, belongs to the State of Florida. No pilfering of treasure is allowed.

But what is found on the beaches is another story. Even what is found accidentally by divers finning along the coast looking perhaps for lobsters and finding instead maybe a solid gold platter is no longer looked upon by state authorities as an unpardonable sin worthy of fines or imprisonment. At least not yet, under existing laws. But beware. Such conditions may be subject to change.

Hopefully, however, the day will never come when sport divers will be legislated against diving down through 30 feet of water off Islamorada and seeing the 100-foot-long exposed bottom timbers and ballast pile of the 255-year-old Spanish treasure ship *San Jose* that perished there in 1733, or be fined for poking around in the shells and rubble near the 19 iron cannon and two anchors 900 feet offshore at the *Rio Mar* wreck site north of Vero Beach. Hopefully, the legislation will be directed toward the big-time operators bent on ripping off the wrecks, taking, for example, the cannon and anchors that are still there. Cannon collectors have already abused the site by lifting most of the 72 guns that the vessel originally carried.

IS ANY TREASURE LEFT TO BE FOUND? Listen to the words of Fort Pierce lineman turned diver/treasure hunter, John Durham: "The Thanksgiving Day storm of eighty-four was the best cut we've had in 20 years. I grabbed my detectors, headed for the beaches and didn't come home for ten days and nights...I did better with gold in less than an hour at Corrigan's (one of the 1715 wreck sites north of Vero Beach) than I've done in all my life. I got nine 8-escudo coins currently valued between $6,000 and $8,000 apiece, and nine other coins...Even the wives were walking along without detectors picking up coins..."

Later in the day of the same storm, diver and TH-er (treasure hunter) Ron Hampton of Tampa, uncovered a priceless gold locket with a glass face, Spanish

Dragon pendant and 11 1/2-foot gold chain found on a beach opposite a 1715 Spanish wreck.

inscription and a miniature ivory statue of a saint inside. Not far away, near the sand dunes, his buddy's detector zeroed him in on a small gold cross. The tiny green stones adorning it turned out to be emeralds.

HOW TO GET STARTED. In the words of successful treasure hunters, the advice is simple: "Read all you can on the subject, get yourself a good waterproof metal detector, then start searching the areas where treasure has been found."

Your local library is a good starting place for books on the subject. You will soon find many focusing their attention on areas of Spanish treasure fleet losses, ranging roughly from south of Cape Canaveral to Sombrero Key in the middle of the Florida Keys. Surprisingly, many of these treasure laden vessels were shattered by hurricanes and their riches dumped into snorkeling depth water. Unfortunately, generally speaking, the clarity of these Atlantic waters has much to be desired. Yet, on the good days, divers are sometimes startled to find themselves looking down on a scattered cannon, piles of river-worn Spanish ballast rock, and in some places, the actual worm-eaten timbers of a Spanish treasure ship.

All of these areas are protected historic sites and divers are requested not to molest them. Treat these sites with the same care you might any other irreplaceable antiquity. Look but don't touch. Besides, precious cargo items may lie scattered just beneath the sands of that scarred reef in the background that may have absorbed much of the fallout items. Be assured that all of these sites have literally been looked over to death by both amateurs and professional treasure hunters. What finds are still being made are chance ones, away from the wrecks. The best bet of all are the beaches. But above or below water, the metal detector is the key to making hidden finds.

Treasure hunting Florida's southeast coast.   Bob Burgess

WHAT DETECTOR IS BEST? Expect to pay anywhere from $500 to $850 for a good underwater metal detector. The reason why the underwater detector is best, even working beaches, is because the best time to search is right after severe storms have scoured the beaches. Conditions will be far from safe for anything but a waterproof detector. The pulse induction types are very popular, very good above or below water and capable of detecting coin-sized objects several feet underground. Some pulse induction units require almost repeated retuning, which is a drawback. Others are automatic and require no retuning.Company addresses can be found in any of several treasure magazines on your newsstands. Units suitable for divers are often advertised in dive magazines.

Again, returning to what the experts advise, get a good book and learn how to use your metal detector. Some units produce a tone signal. How that tone sounds, whether it is short and sharp, clicking or booming, will tell you in advance whether you have found scrap metal, an aluminum pull-tab, or a silver coin—before you even dig it! You can even set the detector to respond only to coins or other valuable finds, rejecting the junk! Such refinement allows the hunter to move rapidly through areas laced with trash and target only the valuable. Other, less sophisticated machines, target everything, requiring that you dig everything to find anything of value.

But that and all the other good techniques are explained in metal detecting books. And keep in mind that you may be targeting Spanish treasure along the beaches; but more than likely your biggest find will be rings, coins, keys and jewelry lost not by the Spanish but by today's beach-going crowds. The treasure they leave behind, lost in the sand, is considerable. (And may heaven smite those who bury crushed beer cans on our beaches!)

Incidentally, make a habit of covering all your holes and keeping all the trash you dig for later disposal. It not only makes for goodwill, but your thoughtfulness will keep others from digging the same trash. Most people practice this automatically and provide us all with more pristine beaches to enjoy.

WHAT ARE THE LEGALITIES OF TREASURE HUNTING ONSHORE OR OFFSHORE? Many divers think that because Florida protects its historic shipwrecks and has certain other areas set aside for contract salvage of treasure wrecks, sport divers are not allowed in these areas. This is not true. Sport divers, with or without metal detectors, are allowed—at least under current laws—to dive on all such historic shipwreck sites. All the State asks is that you don't molest them. Period shipwrecks, beyond Florida's three-mile limit, are commonly visited and investigated by all manner of treasure hunters and sport divers. These are under federal jurisdiction and the same common sense behavior applies—disturb nothing, so the next underwater visitor can enjoy the pleasure of looking over the centuries-old shipwrecks and taking time to enjoy them for the historically important artifacts they are.

In your underwater searches, should you find treasure, the question of ownership will come to mind. Legally, what you have found belongs to the people of Florida. The authorities say you should report it to Florida's Department of Archives and History in Tallahassee. It is questionable whether they may claim it. They are not interested in coins. They may be interested in unique artifacts if only to obtain an archaeological record of the item. They may even offer to purchase same at 50% of its appraised value as they have in the past, and the item will then be displayed in the state museum. The state is not particularly concerned about the sport diver underwater with his metal detector—after all, he may be looking for his wife's wedding ring! It is concerned, however, if a vessel cranks up a blower in a restricted area and goes to work wholesale in the treasure hunting business; or if a diver or divers decide they might like to lift one or more of the many old cannon and anchors sometimes found near the sites. These practices are verboten. And just because you don't see a boat on the horizon, or a Marine Patrol zeroing in on you, or that spy in the sky peeking down on you, don't believe that you can get away with any of it. The sites are under surveillance and penalties for molestation of same are severe. All onshore finds, however, are yours!

WHERE TO LOOK. Beach hunters in the know watch the weather like hawks. When strong northeast gales rip down the Florida Straits, they head for the beaches opposite the shipwrecks of the 1715 fleet, those scattered between Sebastian Inlet and just south of Fort Pierce. What they hope for is a five-foot beach cut by storm waves. When that happens, even wives accompanying their metal-detecting husbands have been seen picking up Spanish silver coins off the beach. TIP: Go early to beat the early birds. Last time I tried it at dawn, a handful of real earlybirds were leaving the beach. They had gone there at 2 a.m. and found several Spanish coins in the dark!

Seven hot spots for beach hunters occur along the barrier island opposite Highway A1A, from just south of Sebastian Inlet to just south of Fort Pierce. The two best places to look are along the beaches opposite the Cabin Wreck and Corrigan's. Since the Sebastian Inlet State Recreation Area just north (entrance) and south (camping) of the bridge offers good overnight facilities close to the northern sites, you may wish to proceed from there.

The first access path 1.1 miles south of Sebastian Inlet Bridge will bring you to the beach opposite the Cabin Wreck where Kip Wagner first found his "Money Beach," and later, the offshore wreck that has produced tons of silver-cob coins from the 1714 period. A "U-shaped" scatter pattern of recovered treasure touches the beach here for about a quarter of a mile. Search in both directions. You may see a contract salvage vessel working the site just beyond the surf, where this treasure galleon came to an end.

The next best site is Corrigan's at 9.7 miles south of the Sebastian Inlet bridge. This 1-1/2 mile long kidney-shaped scatter pattern of recovered treasure touches shore for over one mile. The access point is the Turtle Trail Beach Access with paved parking for 23 compact cars. Walk over the wooden bridge across the dunes and search in both directions. Large numbered signs on the dunes are range markers for the contract salvors working offshore. Gold, silver and fine jewelry have been found here by metal-detector enthusiasts. During storms the gated entry to the Turtle Trail parking may be locked. Beach hunters then park on the grassy shoulder of the highway.

FLORIDA

The Next Best Thing To Being There, Is Reading Florida Scuba News

   If you have plans to make a diving trip to Florida, or if you want to learn why Florida diving is so great, you need to subscribe to *Florida Scuba News*. We provide all the late-breaking news, Area Features, the June Charter Boat Issue, the July Lobster Issue, and the January Dive Shop Listing. Add to this list our monthly Manufacturers News, Resort & Travel News, Calendar, Dive Buddies, Dive Clubs, Ned DeLoach's Fish I.Q. Test, articles and stories written by experienced Florida divers and you'll agree that a subscription to *Florida Scuba News* is a piece of dive equipment you can't do without.

FSN
·FLORIDA SCUBA NEWS·

For your monthly subscription, send form and $12.50 to
Florida Scuba News, 1324 Placid Place, Jacksonville, Florida 32205
(payment must accompany your order) .

Name_____

Address_____

City, State, Zip _____

SPECIES	SIZE LIMIT	CLOSED SEASON	DAILY REC. BAG LIMIT	REMARKS
Amberjack	28" fork	- - -	3	
Black Drum	Not less than 14" or more than 24"	- - -	5	Cannot possess more than one over 24".
Sea Bass	8"	- - -	- - -	All Sea Basses.
Bluefish	10"	- - -	- - -	
Bonefish	18"	- - -	1	Cannot buy or sell. Only one in possession.
Hard Clams	1" thick	- - -	- - -	Requirement for boats and harvest gear. Special license Brevard and Indian River counties.
Cobia (Ling)	33" fork	- - -	2	
Stone Crab	2 3/4" claw	Between May 15 and Oct. 15	- - -	Trapping under permit only. Cannot possess whole crab.
Crawfish	More than 3" carapace	April 1-August 5	24/boat or 6/person whichever is greater	Trapping under license only.
Red Drum	Not less than 18"	March, April, May	1	Cannot buy or sell native redfish.
Flounder	11"	- - -	- - -	
Grouper	20"	- - -	5	Includes Yellowfin, Red, Black, Nassau, Gag, Yellowmouth, Scamp.
Jewfish	- - -	Prohibited Harvest	- - -	
King Makerel	12"	- - -	2	Bag limit in Gulf-Atlantic fishery reduced to 1 when federal waters close all harvest.
Spanish Mackerel	12"	- - -	4	
Oysters	3"	June, July, August	- - -	Special regulations Apalachicola Bay-1 bag per person, boat or vehicle, whichever is less per day.
Permit	- - -	- - -	- - -	No more than 2-20" or larger.
Pompano	10"	- - -	- - -	Prohibits sale greater than 20"
Scallops	- - -	April 1-June30	5 gallons whole or 1/2 gallon meat..	Special regulations St. Joe's Bay.
Red Snapper	13"	- - -	2	Bag limit No more than 10 snappers aggregate
Schoolmaster	10"	- - -	10	
Gray (Mangrove)	10"	- - -	5	
Lane	8"	- - -	- - -	
Vermillion	8"	- - -	- - -	
All Other Snapper	12"	- - -	10 aggregate of all having a bag limit	Includes Blackfin, Cubera, Dog, Mahogany Mutton, Queen, Silk, and Yellowtail.
Snook	24"	Jan., Feb., June July and August	2	Cannot possess more than one over 34". Cannot buy or sell.
Tarpon	- - -	- - -	2	Cannot buy or sell — requires $50 tarpon tag to possess or kill.

FLORIDA'S DIVING LAWS

CORAL; Unlawful to take, possess or destroy sea fans, hard corals or fire corals unless it can be shown by certified invoice that it was imported from a foreign country.

CRAWFISH: Lobster must remain in a whole condition at all times while being transported on or below the waters of the state. No eggbearing females may be taken. Use of grains, spears, grabs, hooks or similar devices prohibited. The molesting, taking or trapping of spiny lobster within the Biscayne Bay Card Sound Crawfish Sanctuary within Dade and Monroe counties is prohibited. Divers are required to have a carapace measuring device in possession and to perform each measurement in the water.

SPORTS FISHERMEN'S SEASON: The last full weekend prior to August 1 (two days only). May not possess more than six crawfish per person on the first day, not more than 12 crawfish per person cumulatively for both days.

HARD CLAMS: Unlawful to take, possess or transport clams on the water from one-half (1/2) hour before official sunrise. Unlawful to use any rake, dredge or other mechanical devices to harvest hard clams in any grass bed.

MARINE MAMMALS: (Manatees or Sea Cows, Porpoises and Whales): Manatees or sea cows, porpoises and whales are endangered species. All marine mammals or parts thereof are protected by both state and federal law. Unlawful to take, kill, injure, annoy or molest a marine mammal. Report any harassment, distress, injury or death of a marine mammal to the nearest Florida Marine Patrol office or call 1-800-342-1821.

MARINE TURTLES, NESTS AND EGGS: Unlawful to take, kill, molest, disturb, harass, mutilate, destroy, cause to be destroyed, sell, offer for sale or transfer any marine turtle, marine turtle nest or marine turtle egg. Marine turtles accidentally caught must be returned to the water alive immediately.

OYSTERS: May only be taken from approved shellfish harvesting areas.

QUEEN CONCH: Unlawful to take or harvest any queen conch from the land or waters or to possess or transport any queen conch so taken or harvested.

SNOOK: Unlawful to take by any means other than hook-and-line. Unlawful to take with live or dead natural bait in conjunction with a treble hook. Snatching prohibited.

STONE CRAB CLAWS: No trapping except under permit from the Department of Natural Resources. Use of spears, grains, grabs, hooks or similar devices which can puncture, crush or injure the crab body is prohibited. Legal claw or claws may be taken but live crab must be released. Unlawful to remove claws from egg-bearing females.

DRUGS-POISONS: Illegal to place drugs or poisons in the marine waters except by permit from the Florida Department of Natural Resources.

EXPLOSIVES: The use of explosives or the discharge of firearms into the waters to kill food fish is prohibited. The landing ashore or possession on the water by any person of any food fish that has been damaged by explosives, or the landing of headless jewfish or grouper, if the grouper is taken for commercial use, is prohibited.

SPEARFISHING: Unlawful to spearfish as follows:

(1) Within 100 yards of all public bathing beaches; commercial or public fishing piers; and that portion of any bridge where public fishing is permitted.

(2) Within 100 feet of that portion of any jetty that is above the sea surface, except along the last 500 yards of the exposed portion of any jetty extending more than 1,500 yards from the shoreline.

(3) For the taking of all species of ornamental reef fishes, i.e., surgeonfish, trumpetfish, angelfish, butterflyfish, porcupine fish, cornet fish, squirrelfish, trunkfish, damselfish, parrotfish, pipefish, sea horses and puffers.

(4) In Collier County and that part of Monroe County from Long Key north to the Dade County line.

(5) In or on any body of water under the jurisdiction of the Division of Recreation and Parks of the Florida Department of Natural Resources. (The possession of spearfishing equipment is prohibited in these areas.)

(6) For the taking of any species whereby the taking by spear is prohibited.

STATE BOUNDARIES: Florida's state waters consist of all waters within nine nautical miles of the shoreline in the Gulf of Mexico and three nautical miles of the shoreline in the Atlantic Ocean. For fisheries purposes, the federal waters are those waters 200 miles seaward of state waters. One nautical mile equals 6076 feet.

Marine Patrol Offices

1. Panama City — Naval Coastal System Center Bldg. 432. (904) 234-0211
2. Carrabelle— South Marine St. (904) 697-3741
3. Homosassa Springs — Highway 19 S. (904) 628-6196
4. Tampa — 5110 Gandy Blvd. (813) 272-2516
5. Fort Myers — 1820 Jackson St. (813) 334-8963
6. Miami — 1275 N.E. 79th St.(305) 325-3346
7. Titusville — 402 Titusville Causeway. (407) 383-2740
8. Jacksonville Beach — 2510 2nd Ave N. (904) 359-6580
9. Marathon — 2835 Overseas Hwy. (305) 743-6542
10. Jupiter — 1300 Marcinski Rd. (407) 626-9995
11. Pensacola — 1101 E. Gregory St. (904) 444-8978

Headquarters

Marjory Stoneman Douglas Building, 3900 Commonwealth Blvd., Tallahassee, FL 32399 (904) 488-5757

RESOURCE ALERT LINE

If you should observe any violation of the marine fishery laws of the state, please call **1-800-DIAL-FMP**

U.S. Coast Guard

Marine & Air Emergency

St. Petersburg...........(813) 896-6187   Mayport (Panhandle)........904) 246-7341
Miami.......................(305) 350-4309   Charleston, SC ....(803) 724-4382
Key West..................(305) 350-5328   At Sea ....VHF 16 or HF 2182

Divers Down Flag

The recognized divers down flag in the US has long been a rectangular flag, with a white diagonal stripe running from the upper left to the lower right corner. However, the federal law states that, when diving from a boat, divers must display the international Alpha flag, a blue and white flag. According to federal law, divers who fail to display the Alpha flag will be deprived of any legal rights or course of action against boaters in the event of an accident.

The Alpha flag is necessary for legal protection, but its meaning is not yet widely known. The traditional red and white divers flag is recognized, but provides no legal protection. Consequently, it is being recommended that divers fly both flags, until such time as the Alpha flag becomes effective in meaning. It should be noted that divers can be cited for failure to display the Alpha flag, and that the flag should not be flown unless diving operations are in progress.

Minimum size of flag(s) flown should be 12 by 12 inches. It should be positioned at a height of at least three feet above the surface. In open water, boaters should stay at least 300 feet from the flag. If you must approach the area of the flag, do so at a very slow speed and watch closely for diver's air bubbles.

Florida Decompression Facilities

It is important that every diver using scuba understand the possibility of contracting decompression sickness, more commonly known as "the bends." In addition, each diver should know the location of the nearest chamber for proper treatment. The phone numbers listed below are those of the facilities in Florida that provide the service of their decompression chambers to the general public.

Pensacola..............................(904) 452-2141
Panama City ....(904) 234-2281 Ext. 370
Gainesville.............................(904) 392-3441
West Palm Beach ....(305) 844-3515
Miami ....(305) 350-7259 or 446-7071
or 445-8926

From 7:30 a.m. to 4:15 p.m., a listing of all chamber locations in the United States may be obtained by calling (512) 536-3278.

In addition, the Diving Accident Network, headquartered at Duke University, is available 24 hours a day for emergency diving accident information, including chamber locations. Contact them in an emergency at (919) 684-8111.

THE LIVING REEF

BY TOM AND WANDA ARTEAGA

One might compare the life cycle of the reef to the well-known story of ancient Troy where the remains of civilization after civilization become the foundation for new coral growth, and life begins again.

In order to understand how a coral reef is formed, one must keep in mind that it is a living animal (Coelentrata) which reproduces both sexually and asexually, and that a single coral animal is called a polyp. Individual polyps vary in size from eight inches to those which can barely be seen by the naked eye. The most common, however, is about one-eighth of an inch long.

A coral reef begins to form when one or more polyps settle on a firm surface and attach to it. The polyps assimilate the calcium which is in the seawater and deposit it beneath and around their bodies. This process forms a "cup" of calcium carbonate, otherwise known as limestone. A particular type of coral is formed when polyps with the same hereditary characteristics come together to form colonies of varying sizes and innumerable shapes and colors. This phenomenon takes place mostly on the warmer western sides of oceans and in open tropical seas. Southern Florida is the only place in the continental United States with water warm enough for coral reefs to subsist.

Coral polyps feed on planktonic forms of life which move with the tides and currents. The coral communities are most active during changing tides when there is a better chance for the polyps to catch food with the tentacles that surround their mouths. When there are sufficient ocean currents to furnish food and the water is warm, relatively shallow and clean, there is a good possibility of a reef evolving.

Living coral grows at the rate of 1.2 centimeters a year—less than one-half inch. Even this slow growth can be inhibited. The tiny polyps are easily clogged and killed by silty or polluted water. Improper sewage disposal and dredging are two of the reef's worst enemies.

As divers who know, understand and love the Florida reefs, much of the responsibility for protecting them should be ours. If we do not fight for and support proper legislation to save them, these magnificent ecological systems called "living reefs" will become underwater deserts.

There are many things individual divers can do to protect the reefs on each visit. Start by seeing that the dive boat is anchored properly. The thousands of anchors used annually on reefs can cause tremendous damage. If mooring buoys are available, use them instead of anchoring. Always anchor on the edge of the reef in the sandy areas. Never throw the anchor, but lower it gently to the bottom. Grappling type anchors seem to cause less damage to the reef than heavier types.

It is common on shallow reefs to see snorkelers resting on coral heads. Each time this flagrant practice occurs thousands of the tiny polyps are destroyed. Take pride in your buoyance control so that your diving skills will help you avoid contact with the reef.

Fewer sights are more disheartening than a diver tearing through living coral to obtain a particular shell specimen or collector's item. The most caring, lasting and beautiful underwater souvenirs are taken on film.

Cans and other trash are no enhancement to the lovely corals. It takes only a few moments to collect litter and take it back to the dive boat. The most foresighted divers always show their care in this small but exceedingly important way.

A privilege, a responsibility and a joy, our reefs are a living celebration of color and drama that lure and enchant us. May we continue to share and delight in them.

FISH IDENTIFICATION
A New Breed of Underwater Hunter

With environmental awareness marking the decade of the nineties, many divers are taking up a new and rewarding underwater activity — fish identification. Much like their terrestrial counterparts, bird watchers, who spend their time scouring the countryside in the hope of sighting rare species, fish watchers have taken their search to the reefs. Now, instead of passively swimming through the undersea environment, the new breed of underwater hunters actively seek out and learn to identify the dozens of fishes that surround them on every dive. It is rapidly becoming a competitive activity between divers to see who can spot and identify the most species on a single dive or during a dive vacation. Like spearfishing, the passion of the hunt is still there, but the damaging spear has been replaced with a pencil and slate. In the future, divers will, hopefully, be known for the number of fishes they identify instead of what they kill.

Making the efforts of fish watchers even more purposeful is their ability to take fish species surveys in areas where they dive. Currently there is no effective means of substantiating declining marine life on our reefs — information essential for environmental lobbies. For the first time, this important data is being compiled by the Reef Environmental Education Foundation. REEF is a non-profit organization of active divers dedicated to the preservation of marine life and habitats through research and education. There are simply not enough marine biologists working in the field to assemble this volume of information. Recreational divers form the only group large enough, active enough, or concerned enough to regularly monitor the vitality of our reefs. With their help, using basic identification skills, a data bank documenting the geological diversity and population trends on and around our reefs in various geographical locations can be built and maintained. If divers fail to meet this challenge, the prospects of these fragile marine habitats flourishing into the twenty-first century will be diminished.

REEF is enlisting divers interested is taking field surveys on the reefs of Florida, the Caribbean and Bahamas as part of their Reef Fish Identification Project. Advanced identification skills are not necessary. Even beginning fish watchers can make significant contributions. As the Project grows and the skills of participants improve, more sophisticated data can be gathered and sighting information can grow to include corals and other marine invertebrates.

Many dive retailers, resorts and live-aboards are helping to train divers for the Project by teaching fish identification specialty courses and scheduling special identification dives. To become a member of REEF and receive an information packet about participation in the Reef Fish Identification Project, contact your

local dive retailer or write REEF at Dept. W, 1861 Cornell Road, Jacksonville, FL 32207. There are no dues or fees and your name will not be supplied to other organizations. A list of dive stores and operators presently offering specialty instruction in fish identification is also available upon request.

FLORIDA LOBSTER

BY ED DAVIDSON

The Florida lobster is actually the crawfish, Langusta. It lacks the pincer claws characteristic of the true northern lobster. Its taste, when broiled in butter, however, is just as delightful as that of the lobster. Langusta typically crawl out in the open in search of food just after sunset, and return to their homes under rocks, ledges and coral heads at predawn light. They are basically scavengers which augment their diet with small fish and crustaceans. There are also periodic mass migrations, or crawls, which may involve thousands of lobster in weaving columns, miles long. Such migrations have occurred in the months of September, December and late March, under circumstances which are still not clearly understood.

Lobster season is open from August 1 to March 31. It is closed during June and July, the months when spawning and molting cycles reach their peak. Commercial lobstermen use slat traps baited with a variety of tasty tidbits ranging from kerosene-soaked bricks, sardines and cat food to cowhide and chopped fish heads.

Without a commercial license, you can take lobster only by hand or with hand nets. They must never be gigged or speared. The minimum size crawfish that can be kept must measure three inches from the apex of the eyes between the supraorbital horns to the after edge of the thoracic shell or carapace. It is a state law that divers in possession of lobster have a gauge for measuring this dimension. Possession of a "short" can cost you a hefty chunk of vacation money and a visit to the local magistrate. The lobster laws are actively enforced by the Florida Marine Patrol, which maintains a constant presence with fast patrol boats and metes out substantial fines for violations such as possession of "shorts"; spearing and gigging; having more than the legal limit of 24 lobster in your boat; wringing off tails before you reach the dock; possessing egg-bearing females (easily identified by the yellow eggs covering their undersides); and robbing commercial traps.

Since the local lobstermen have been known to occasionally unlimber their blunderbusses in the direction of people poaching their traps, it is best to stick to grabbing the desired crawfish by hand with a good pair of gloves—then fire up the broiler!

MANATEES

BY SHARELL HOLVERSON

In addition to Florida's many well-known attractions, each year hundreds of divers enjoy a special treat—swimming with the rare and endangered manatees. The King's Bay area of Crystal River is the most accessible spot for divers to encounter these animals in their natural habitat. The experience is especially unique in light of current estimates which indicate there are only 750 to 1,000 sea cows in Florida waters.

Manatees, like people, are mammals and need to maintain a constant body temperature. Though they live in water all their lives, they breathe air at the water's surface through nostril flaps. Females suckle their young, which are born alive. Births are usually single, but twin births are seen.

The huge grey mammals are lured to spring-fed freshwater rivers by their constant 72-degree temperatures during the cold winter months. They are also seen near warm water discharges of power plants along the coast when water temperatures fall.

The 'composite' animals have skins resembling that of elephants, grayish in color, varying from rough to smooth in texture, and with occasional sparse body hairs. Flippers, which appear to be attached backwards, have nails reminiscent of an elephant's toenails. Their noses are short, enhancing a 'puppy dog' appearance. Rubbery, bristle-covered jowls easily manipulate the manatees' favored food supply—hydrilla. Although relished by manatees, the frilly aquarium grass is considered a nuisance to boaters. Sea cows have only grinding teeth. No incisors or cutting teeth exist in the front of their mouths—only bony arches or 'gums' similar to those of a cow. They are incapable of attack, flight being their only defense.

Though shy by nature, the Crystal River manatees have become accustomed to snorkelers over the years. Some individual animals actually seek out and enjoy the attention of nonaggressive divers.

As a snorkeler or diver quietly enters the water from his boat, he will often find a large gray shape within a few feet, observing him. A moment of mutual respect may set the stage for a memory of a lifetime. As a diver hangs motionless in the water or swims slowly, the curious animal may invite contact.

At times, the manatees enjoy gentle, openhanded stroking on their torpedo-shaped bodies and short necks, but rarely on their large, flat tails. Finger-jabbing is never welcomed! As they become familiar with a diver, face and flipper stroking seems to be a pleasure to them. A respectful diver may soon have a huge sea cow rolling over in the water to expose his entire lighter-colored underside for petting and stroking. Some will even emit audible grunts of pleasure.

High-pitched squeaks and whistles are also heard in the water near manatees. Although not anatomically the same as the echolocation used by dolphins, manatee cows and calves keep track of each other in murky waters with their sounds. They also have been observed to whistle when confused or frightened by approaching boats or aggressive divers. Do not chase or pursue manatees! Try not to disturb a sleeping, eating or nursing sea cow. An animal resting on the bottom or hanging motionless near the surface may be sleeping. He will come up periodically for air even in his sleep and may be startled by a diver he didn't know was there.

In addition to hydrilla, manatees also eat water hyacinths and other river and sea grasses. They may consume as much as 100 pounds of weeds and grasses a day to maintain their impressive body weight of up to 1,500 pounds. Obviously, this much eating takes a lot of undisturbed time. Even in poor visibility, it is easy to detect when a manatee is eating. Loud crunching, grinding sounds travel for some distance underwater.

Nursing behavior is often seen. The mammary glands of a female are just under the flippers. A baby manatee at his mother's 'elbow' is a good "Do Not Disturb" sign.

Manatees are often deeply cut by boat propellers. It is important to recognize the signs of their presence from above the surface of the water. Periodically, they surface to breathe, sucking in air with a loud, rushing sigh. Alert ears will hear them easily, especially if boat motors are at idle speed (required by law in manatee waters from November 15 through March 31). Also, watchful eyes will note evenly spaced swirls in the water several feet apart made by a manatee's powerful tail as he swims near the surface.

SEA TURTLES

BY SUSAN KUNTZ AND SHARELL HOLVERSON

Modern sea turtles evolved into their present shelled and toothless form over 100 million years ago. No other air-breathing vertebrate group has lasted unchanged so long. But during the last 100 years, man has managed to alter their evolutionary course decisively. The body armor (or carapace) which has well protected them for so long has done little to shield sea turtles from devastation by man. Once prolific in tropical waters, today, sadly, all varieties of marine turtles appear on the endangered species list. Those most valued for meat, eggs and shells closely brush the limits of extinction. The green turtle has been especially desired for its meat and eggs, while the hawksbill has been prized for tortoise shell, used in making ornaments. Sea turtles are further pressed by pollution and man's rapid encroachment onto once undisturbed beaches, mandatory for successful egg-laying.

Sea turtles are not able to withdraw into their shells for protection as many land turtles do. They are dependent on their size and swimming abilities for defense. Impressive in terms of sheer body mass, some leatherbacks are estimated to have reached 1,500 pounds and eight feet in carapace length. But the most

remarkable capability of the sea turtles is their graceful flight underwater. They do not paddle, but "fly" like marine birds. Some are able to skim over 100 yards of reef in ten seconds.

In Florida's coastal waters, sea turtles mate in April, May and June. Nesting season is May through August. Females come on land only to lay their eggs. The mothers drag themselves above the high-water mark and dig a small hole with their hind flippers. After laying 100 or more eggs, they fill in the hole and then, using their front flippers, disturb a large area of sand so that the location of the nest is not apparent. Man, seldom fooled by this effort at camouflage, often decimates the nest, removing the eggs for food.

The babies, which measure only a little more than an inch, face a very difficult beginning in life. When they hatch at the end of the summer, they must make a precarious trek across open sand to their saltwater home. Lying in wait are sea birds, crabs, dogs, cats, raccoons and other animals and insects which relentlessly pursue their vulnerable prey. During this difficult period, only one out of 100 baby turtles is destined to survive. Even when they reach the water, large fish and sharks continue the unremitting pursuit.

It is against the law to harass an endangered species, including all marine turtles! Even riding a turtle underwater is considered harassment. It is also illegal to bring any turtle product into the U.S., including captive-bred turtles. Products made from turtles, often purchased as souvenirs, are confiscated by U.S. Customs.

There are several things divers can do to help sea turtles, in addition to obeying the laws.

1) Be careful boating. Marine turtles spend much time basking at the surface. Every year turtles die from propeller injuries.

2) Learn to identify turtle tracks and nests. If you see a nest that looks as if it will be inundated by high tide, report it to local authorities or turtle specialists. If you help in removing the eggs (with permission), be careful to move one egg at a time. Do not rotate it. Set it down in the new nest or bucket maintaining the same vertical axis.

3) If you see or suspect any egg poachers, report them immediately to the local police or the Florida Marine Patrol.

4) Do not disturb a mother turtle on her egg-laying mission. She may return to sea and abort her efforts altogether.

5) If you spot the tiny tracks of freshly hatched turtles, follow the trail to be sure the babies made it safely to the sea. Upon hatching, the young turtles are instinctively drawn toward the light of the horizon. However, they often become disoriented by city and traffic lights and head away from their intended destination. They may also become entangled in land vegetation or sargassum weed. Deep impressions in the sand and seawalls create other barriers.

Divers who have observed the grace and beauty of the armored knights of the reef will long carry the image of these unique beings. Unfortunately, their continued survival is at risk. Only concern and effort will halt their rapid course toward extinction.

JEWFISH

One of Florida's natural phenomena is the large jewfish colonies that inhabit the ledges and wrecks of her Gulf waters. Dropping down through 50 feet of haze onto a wreck is exciting, but to suddenly discover 15 to 20 living monoliths moving boldly through the broken rubble is awesome. Often weighing several hundred pounds (some reported 700 pounds), the huge fish are lords and masters of their domain. Their formidable size deters all natural enemies and allows them to live unmolested for years.

Jewfish are bottom-dwellers, usually spending their lives in protected areas at shallow to moderate depths. Although cumbersome in appearance, the fish can cover short distances quite quickly with a stroke of their powerful tails. They draw fish and crustaceans into their voluminous gullets by suction created when they open their cavernous mouths. Dinner is held tenaciously by thousands of small, rasp-like teeth that cover the jaws, tongue and palate. The prey is then swallowed whole.

There are few Gulf divers who do not have a favorite jewfish story to relate. These tales most often involve the fish's gluttonous, assertive habits. Curious by nature and emboldened by size, they occasionally approach visitors who enter their realm. Often, when irritated, they make a loud "booming" sound with their gills, warning those who approach of their pugnacious mood. Although seldom aggressive toward divers, they have been known to attack a spearfisherman's catch and, intent on stealing an easy lunch, to engulf a finned leg in the melee. Such encounters are never fatal, but the leg is seldom pulled free without a nasty set of lacerations.

A few divers choose to make pets of individual jewfish. A friendship is nurtured by repeatedly taking cut bait on dives to a favorite underwater site. A voracious appetite generally proves more overpowering than caution. Patience and gentleness soon have the big fish coming up to greet divers as they descend toward its lair. This practice is rewarding to divers, but often detrimental to the longevity of the fish. The befriended giant becomes an innocent victim to men with spears.

Because of their large trophy-size and fearlessness, jewfish have already become void from most readily accessible Gulf diving locations. This devastation has come about by a disregard for their uniqueness. Their stringy, usually wormy meat is seldom palatable. A spectacular sea creature that could thrill hundreds of underwater observers is destroyed simply to inflate a single hunter's ego.

SPEARFISHING

Skin diving started with spearfishing. Selecting, stalking and subduing quarry in an unknown environment, without the ability to draw a second breath, was a challenge well-suited to man's instinct for adventure. Throughout the '50s and into the 60s, few ventured below without a spear in hand. Divers were known by the amount and size of fish they brought to the surface. Few of the enthusiastic hunters imagined that their newly discovered sport would have any negative impact on reefs that, then, overflowed with fish.

A new generation of divers appeared when scuba became available to the public. This group soon far outnumbered the handful of free divers that preceded it. Now able to spend long periods underwater, many divers became devoted observers of reef systems. Through their observations, one unequivocal reality became apparent—where reefs had been regularly hunted, few, if any, game fish remained. Word of the depleted reefs was received by a world just awakening to its obligation to protect delicate environments from man's rapid encroachment. What followed came to be known as "the great spearfishing controversy."

For over three years, underwater ecologists and underwater hunters fenced in the arena of opinion. Never before or since has the generally composed diving community rippled with so much passion over a single issue. Ecologists, armed with fish-barren reefs and the trend of the time, forged an unrelenting attack on those who dared to carry a spear. Such august personalities as Dr. Hans Haas and Philipe Cousteau gave credence to the inspired lobby.

Two basic arguments were leveled: spearfishing had a negative impact on the ecosystem of the reef, and it wasn't equitable for a few hunters to deny many divers the right to observe the reefs complemented by nature's full array of inhabitants. Soon sanctions were being instituted against the hunters. The diving equipment manufacturers were discontinuing the sales of spearguns. Resort islands banned spearing; tournaments were cancelled; and charter boat operators ceased to allow guns on board their vessels.

The old guard of spearfishermen, who had so recently been favored with esteem, watched stunned as they were suddenly cast in the role of villains. They rallied with a salvo of their own. A primary contention was that commercial and line fishermen annually took many more fish than the spearfishermen and that neither group could be selective in what they took from the sea.

Another accusing finger was pointed at novice divers and meat hunters. Beginning hunters, in haste to prove their prowess, shot indiscriminately, killing everything from sea cucumbers to queen angels. To produce profits at the fish market, meat hunters systematically annihilated all game fish from entire areas. It was stressed that true underwater hunters were not the culprits and should not be denied the right to procure fish for their tables.

Nearly a decade has passed since the controversy settled, but in its wake has come constructive change. As with most disputes, resolve tends to fall somewhere between the ideologies of both camps. For a time, diving ecologists seemed a bit overzealous in their meritable attempts to protect the seas. A ban on spearfishing would have lamentably denied many sportsmen the right to savor one of hunting's most challenging and rewarding experiences. But, at the same time, veteran spearfishermen were well aware that many claims made against their sport were valid. They, too, had observed a marked decline in gamefish activity in repeatedly hunted areas.

One fact stood solidly: the voiding of game fish on reef systems and wrecks frequented every year by hundreds of sport divers would no longer be tolerated. Changes had to be made in order to protect fish life on popular dive sites and to preserve the reward of underwater hunting.

The experienced spearfishing contingent tacitly imposed a set of standards that has improved spearfishing's image and also helped renew fish life at many prominent diving spots. Today, spearguns are less often seen at shallow, close-in, clearwater sites. The skilled hunters, instead, voyage farther out to sea, relentlessly searching for new bottom, often in deeper, darker, less placid waters. They limit the hunt to fresh meat for their dinner tables. Large trophy size fish are genetically superior and far more prolific; and, therefore should be passed over for smaller fish.

The education of new divers is paramount to preservation. First, they must be taught that the world's waters are not theirs to greedily plunder. Instead, they must learn to understand, respect and help sustain the seas' vulnerable creatures.

Paul Humann

FLORIDA KEYS FACTS

The Florida Keys are a string of more than 200 islands that extend the length of the Florida peninsula for 180 miles to the sea. From Jewfish Creek to Key West, the island chain is connected by the longest overseas highway in the world—a 108-mile span that links 31 islands with 40 bridges.

The islands separate the shallow flats of the Gulf of Mexico from the Florida reef that lies on the edge of the Gulf Stream in the Atlantic. The reef, which parallels the entire length of the Keys, is the only living coral reef on the North American continent. It is the home of one of the ocean's greatest ecological systems.

Where To Dive

Often divers who travel to the Keys for the first time are under the illusion that they can experience good diving from the shore. Unfortunately, such opportunities are nonexistent throughout the islands.

Mangrove thickets cover much of the shoreline on both the Atlantic and Gulf sides. Here the water is turbid and dark, offering scarcely one to two feet of visibility. Extending far out from the mangroves are thick turtle grass, meadows and mud flats. Low-profile patch reefs begin two to three miles out, but are scattered and unimpressive.

Diving under bridges and in canals is hazardous and often illegal. A strong tidal flux is almost continuous in such areas. Such currents can, in a matter of minutes, quickly sweep unsuspecting divers far away from their entry point. Beach diving is also disappointing. The few beaches to be found have shallow barren sand beds running out to sea.

The premier diving grounds that are so famous and attract thousands of divers each year are the living coral gardens located on the Atlantic side of the island chain. These impressive marine habitats are all five to six miles from shore, requiring a boat for access.

Diving boat services are the most practical way to visit these dramatic reef systems. Dozens of safe, well-maintained boats designed specifically for diving are available for charter throughout the Keys. The captains know the waters and what the diver wants. They operate their reef trips to see that you get the most out of your vacation. Two-tank morning dives leave the docks between 8 and 9 a.m. Afternoon excursions depart around 1 p.m. When the seas are calm, exciting night dives are scheduled. Reservations are recommended during holidays and on weekends.

Climate

Mild, subtropical weather has long been a trademark of the Keys. The Gulf Stream's warm waters that sweep the Atlantic side of the islands have a moderating effect on the climate. Trade winds that rise from these waters as they move through the Florida Straits temper the winter's air and cool the summer's heat. Air temperature averages just over 77 degrees F. November through February are the coolest months, when nighttime temperatures occasionally drop into the upper 40s. However, the morning's sun will generally raise such temperatures to a comfortable 70-degree range by early afternoon. In the summer, daytime temperatures hover in the upper 80s, slipping into the 90s in late July and August. During most of the day and early evening light, variable southeastern sea breezes bring relief from the heat.

In the summer, thunderstorms can be expected. During the heat of the day, squalls move in from the Florida Straits leaving the afternoon still and muggy.

Fowley Light

Richard Collins

Water temperatures on the outer reefs fluctuate little winter to summer. From November to March, a light wetsuit is required when the water drops into the lower 70s. In the summer, the sea is as warm as a tepid bath. Deep-water divers may occasionally pass through thermoclines caused by a spinoff eddy that upwells cooler water from the deep stream.

Heavy winds are the diving boat's nemesis. Prevailing breezes blow from the east-southeast. Strong easterlies play havoc with the ocean's surface. In late winter and April, blows from 10-15 miles per hour sometimes occur. These can raise a 4- to 5-foot sea that makes diving uncomfortable and lowers the water visibility considerably. During the summer months, when the wind is light, the ocean lies smooth, picking up brief chops as afternoon thunderstorms approach.

June through October are the hurricane months. On the average, one will strike the Keys every seven years. These severe storms that once struck the great Spanish Plata Flotilla without warning are now forecast days in advance, allowing inhabitants and visitors ample warning.

The Overseas Highway

Driving The Overseas Highway

Most divers drive to the Keys. The ability to drive from island to island is one of the area's compelling attractions. However, caution is the rule of the day when driving the Overseas Highway. Traffic is often quite heavy on US 1 and generally moves slowly. Recreational vehicles and boats in tow are everywhere along the route. Give yourself plenty of time to make it from one destination to the next; relax, don't drink, and be patient. Everything seems to move more slowly in the islands. Most of the highway and bridges are now four lanes, but traffic is still restricted to two lanes in many areas.

The 25 miles from Florida City at the end of the Turnpike Extension to Key Largo are particularly dangerous. Most of this drive is two lanes bordered by soft shoulders. There are three passing zones on this stretch. Wait for these areas before passing. When driving down, avoid the heavy traffic on Friday afternoon and evening, and again during your return, on Sunday afternoon and night.

Small green mile markers are posted along US 1, beginning south of Florida City and ending in Key West, 127 miles later. These are handy reference points that are commonly used by local businesses to help direct you to their door.

Lenghth of Bridges
of the Oversees Highway

Jewfish Creek ....223 ft.	Vaca Cut ....120 ft.	Park...............................779 ft.
Key Largo Cut ....360 ft.	7-Mile.....................35,716 ft.	North Harris.................390 ft.
Tavernier Creek.............133 ft.	Little Duck-Missouri ..800 ft.	Harris Gap.....................37 ft.
Snake Creek ....192 ft.	Missouri, Ohio ....1,394 ft.	Harris...........................390 ft.
Whale Harbor ....616 ft.	Ohio,Bahia Honda...1,005 ft.	Lower Sugarloaf.......1,210 ft.
Tea Table Relief ....226 ft.	Bahia Honda ....5,356 ft.	Saddle Bunch 2............554 ft.
Tea Table ....614 ft.	Spanish Harbor........3,311 ft.	Saddle Bunch 3............656 ft.
Indian Key ....2,004 ft.	North Pine ....620 ft.	Saddle Bunch 4............800 ft.
Lignum Vitae.................790 ft.	South Pine ....806 ft.	Saddle Bunch 5............800 ft.
Channel 2 ....1,720 ft.	Torch Key Viaduct......779 ft.	Shark Channel..........1,989 ft.
Channel 5 ....4,516 ft.	Torch-Ramrod.............615 ft.	Rockland Channel....1,230 ft.
Long Key....................11,960 ft.	Niles Channel ....4,433 ft.	Boca Chica ....2,573 ft.
Tom's Harbor 3............1,209 ft.	Kemp's Channel ....992 ft.	Stock Island ....360 ft.
Tom's Harbor 4............1,395 ft.	Bow Channel...........1,302 ft.	Key West ....159 ft.

Public Boat Ramps in the Keys

CROSS KEY	Little Blackwater Sound, Bayside
TAVERNIER	Harry Harris Park, Oceanside
	Tavernier Creek at Bridge
BAHIA HONDA KEY......................	Oceanside
BIG PINE KEY	Bahia Honda State Park, Bayside
LITTLE TORCH KEY......................	Old Wooden Bridge Marina
WEST SUMMERLAND KEY.	Bayside
CUDJOE KEY	Little Torch Marina, Bayside
	Mid-key, Bayside
	Bow Channel, Bayside

Islands Crossed
by the Overseas Highway

Key Largo	Grassy Key	Bahia Honda Key	Saddlebunch Keys
Plantation Key	Crawl Key	West Summerland Key	Big Coppit Key
Windley Key	Fat Deer Key	Big Pine Key	Rockland Key
Upper Matecumbe	Vaca Key	Little Torch Key	Boca Chica Key
Lower Matecumbe	Knight Key	Middle Torch Key	Raccoon Key
Fiesta Key	Pigeon Key	Ramrod Key	Stock Island
Long Key	Little Duck Key	Summerland Key	Key West
Conch Keys	Missouri Keys	Cudjoe Key	
Duck Keys	Sunshine Key	Sugar Loaf Key	

POOLS IN THE FOREST
An overview of Florida's spring diving

No one ever quite knows what to expect from a first visit to the Florida springs. This is understandable because each of the more than 120 diveable springs, sinks and clearwater rivers found scattered in the State's northern and central woodlands is a unique creation with its own character and captivating mystique. Their common bond is cool, fresh, clear water—the singular feature that sets them apart and defines each as a natural wonder.

The rare water comes from a network of underground rivers that weaves like a web inside Florida's extensive bed of Suwannee limestone. The hidden rivers are replenished by rainwater that slowly trickles through the porous rock until distilled to perfection. Only then does it join the aquifer—a moving reservoir, crystal pure.

Springs are formed when this groundwater breaks to the surface through a geological weakness. Where this occurs, clear pools form. The overflow is channeled into surface streams, called runs, that wind along a path of least resistance toward a nearby river. A run's size depends on the amount of water feeding the basin. Some are so large that they are used as navigable waterways.

Occasionally the earth collapses into the aquifer, forming sinks. These deep openings fill but do not overflow like springs. Instead, the water feeding into the basin channels back into the ground through a downstream opening called a siphon.

Because of their varied sizes, shapes and settings, springs and sinks are difficult to describe. The only way to comprehend their uncommon beauty is to make a personal trek south to Florida's spring country and see them for yourself.

Such an expedition is a simple matter requiring little more than basic scuba or snorkeling equipment, a guidebook and a flair for adventure. Most springs are located in the backwoods, far from Wendy's Burger and clear TV reception. It is a great place to get away, unwind, relax and enjoy an unspoiled piece of our earth. A trip through a beautiful southern hardwood forest is an inexpensive, enjoyable family vacation. There are plenty of activities to keep the family busy. Sightseeing, camping, swimming, fishing, canoeing and tubing down clearwater runs can be enjoyed by all.

Unfortunately, a widespread misconception about spring diving has lingered for years. Many sport divers and instructors alike still confuse the term spring diving with cave diving. Accordingly, many still associate the spring pools with danger.

In reality, a spring basin offers the safest diving imaginable. The freshwater pools have no currents or great depths and are unaffected by sudden weather changes. The undeserved reputation for danger comes from the occasional drownings that occur inside the caves that feed the pools.

Each year a few untrained cave divers venture into the tunnels, beyond sunlight's glow. Suddenly, the world's safest diving turns into an extremely perilous situation. The unwary divers fail to realize the hazard created by a stone ceiling. If an emergency occurs, escape is not simply up, but an exact retracing of the route taken during the swim into the dark chamber. Over the years the number of such unnecessary tragedies has declined. Education is the key. Two cave diving training organizations, the National Speleological Society—Cave Diving Section

Devil's Eye at Ginnie Springs.

(NSSCDS) and the National Association of Cave Divers (NACD), work, diligently to warn sport divers about the dangers of underwater caves. Their on-going efforts have been a great service to our sport. Neither group solicits students, but, if a diver is determined to explore the caves, both provide proper training.

Those open water divers who are troubled by confined spaces should feel safe exploring the first cavern room in the larger caves. Stay within sight of the entrance light and there is little danger. Leave underwater lights on the surface; this prevents being enticed into the tunnel's darkness. Looking out from a cave entrance on a blue, sunlit pool is one of the underwater world's most dramatic vistas. It alone is worth a trip inside.

Diving trips to the springs can be planned for any season. In the warm summer months the cool 72-78 degree water gives a refreshing lift; in the winter the unfluctuating temperature feels warm as a bath. However, after periods of prolonged rainfall many spring areas are inundated by black water from the swollen rivers. Such flooding can last for weeks. It is wise to call ahead before planning your visit.

Most spring basins are 10 to 20 feet at the vent and taper up to shallow water at the pool's edge. A few have depths from 40 to 80 feet, while Forty Fathom Grotto near Ocala plunges to 240 feet. Snorkeling is excellent at all sites. This is the best way to get a firsthand view of the active fish communities that spend most of their day hidden near grass beds or under lily pad fields.

Diveable spring sites can be divided into four groups: wild springs, commercial diving parks, county and state parks, and tourist attractions. The largest number of forest pools still remain wild—not situated in a developed area or commercially operated. You simply drive to these basins and do as you like. There are no rules, no hassles and no time schedules; you are on your own.

Commercially operated diving parks are becoming more common each year. Although you have to pay an entrance fee and follow a few guidelines, the services provided are well worth the price. They have clean restrooms with showers, campgrounds, air stations, canoe rentals and small concession stands.

Several springs are located on county or state park grounds. County facilities are usually closed to scuba but allow snorkeling. They are used by local residents for swimming and picnicking. The State Park Service now maintains a dozen spring areas. It has done an outstanding job protecting the springs' fragile environment, and, at the same time, allowing divers the freedom to enjoy the basins.

Category four are tourist attractions. These are large commercial ventures that advertise on billboards along the highways. Scuba and free diving have been replaced by glass-bottom boats and underwater viewing windows.

The State's diveable springs are concentrated in four geographic regions: north central Florida, along the banks of the Suwannee and Santa Fe Rivers; throughout the Panhandle, from Pensacola to south of Tallahassee; in the Ocala National Forest; and along the Gulf Coast as far south as Tampa.

On the Suwannee's eastern bank, just past the railroad tracks, sits the small farming community of Branford. When traveling Florida's backroads, it would be easy to pass through its one blinking light without noticing that the tiny metropolis is any different from hundreds of other similar towns in the rural south. But if a stop is made at the Steamboat Motel, the visitor will discover that Branford is host to an alien society—the American sport diver. Miles from the sea, inside the small cafe, the talk around the tables is of tanks and regulators, and on the walls hang dozens of large underwater photographs.

This is the heart of spring diving country. For over 20 years Branford has been the staging area for expeditions into the surrounding countryside where underwater explorers experience the most dramatic freshwater diving in the world.

Four miles north, on the river's eastern bank, the springs begin. The first is Little River, where a forceful flow of clear water erupts from a shallow cave entrance, filling a basin 200 feet across. Further upstream are Royal and Running Springs, then Peacock Slough—a system of eight springs and sinks interconnected by over 21,000 feet of explored passages that recently became a state park.

Unfortunately, the slough area is the first to flood when the Suwannee rises; fortunately, this fact has kept greedy land speculators at bay. Today, the unique spring complex and woods of oak and cypress remain as primitive and pristine as they were 100 years ago. Up the river from Peacock, on the western bank, is Troy Spring. This colossal basin is nearly 300 feet across with depths to 80 feet. Upsteam is River Rendezvous, an inland diving park with a lodge and campgrounds. In front of the lodge near the river bank is small Convict Spring.

East from Branford is the Ichetucknee, the state's most famous clearwater river. Over a dozen springs feed a flowing stream that carries divers on a three mile voyage from the headspring to the US 27 bridge. Below water is white sand, aquatic grass, fish and sholls, while the surface mirrors the ancient forest above.

Not to be missed is Ginnie Springs, found just southeast from the Ichetucknee, on the bank of the Santa Fe River. This is America's most spectacular inland diving park. Years ago when the resort opened, old-time divers grumbled about paying to enter a spring they had dived free for years. Today, no such complaints are heard. The owner, Bob Wray, his family and staff have developed the woodland tract to near perfection. Every necessary convenience is available to the diver and camper, but none of the changes has come at the expense of the environment. If anything, Ginnie, her sister springs and the surrounding forest are healthier and more beautiful than they were 20 years ago.

Florida's Panhandle is a 200 mile strip of virgin forest—a wild tangle of pine and oak crisscrossed by cypress-lined rivers and spring-fed creeks. In the thick of the forest are three superb diving parks.

With a 200-foot basin and depths to 50 feet, Vortex Spring is an ideal dive. A lodge, restaurant, campground and dive shop have been built adjacent to the water. Morrison Springs exemplifies Florida. Its spring pool is lined with a towering stand of moss-covered cypress. Ten miles east is Cypress Spring with a half mile run to Holmes Creek. This is an excellent spot to spend the day.

Going east, toward Tallahassee, there are over ten more springs to visit before arriving at the deep basins, Big Dismal, West Hole and Emerald Sink.

In central Florida, near the Ocala National Forest, is another collection of diveable springs. The state run parks of Alexander Springs and Blue Springs at Orange City are the best for scuba. Just south of Gainesville is Blue Grotto, a large cavern sink recently opened for diving. The cave entrance is lit by floodlights, and concrete steps and a floating dock have been added to make diving a snap. Just eight miles south of Ocala on US 441 is the state's most recent commercial diving park, Paradise Springs. The small fossil-rich sink was used as a local dump before the new owners cleaned and renovated the site and opened it for cavern diving.

The West Coast town of Crystal River was built around King's Bay—a clear body of water connected to the sea. In the winter, when temperatures drop, Florida's largest herd of manatees take refuge in this warmer spring-fed bay. In the last few years nothing has captured the diver's imagination more than swimming with these loveable giants.

In a time when change is rapidly altering the world's fragile underwater environment, Florida's springs remain virtually undamaged. Today, the spectacular basins are more accessible and easier to dive than ever before. Much of the credit for their preservation must be given to the Florida Department of Natural Resources. They use great foresight in caring for their treasures in the forest—the spring pools of Florida.

FLORIDA SPRINGS FACTS

Florida alone accounts for 17 of the 75 "first magnitude" springs in the U.S.A. "First magnitude" spring discharges a minimum of 100 cubic feet of water a second. The state also has 49 springs of the second magnitude, with a flow of between 10 and 100 second-feet. Silver Springs, located northeast of Ocala, is the world's largest spring, with an average flow of 500 million gallons a day (m.g.d.). Wakulla Spring, located south of Tallahassee, is the state's deepest, with a depth of 185 feet inside the first cavern. Its dark passages run through the limestone at over 300 feet. Many of the spring's underwater caves have been explored for miles, with their entire system of intertwining passages running close to the surface. Underground channels often interconnect with neighborhood springs.

The water that pours from underground is cool, varying between 68 and 78 degrees in different springs. These temperatures reflect the mean surface temperatures where the system occurs.

The flow from the springs varies with the amount of water held in their contributing underground basins. Since the basin depends mainly on rainfall for its supply, the discharge of the springs relates to the amount of rainfall over their watershed. If rainfall is low, the spring's flow tempers. This often negatively affects water visibility. Under extreme conditions flows have been known to reverse, sucking surface water underground. In late summer when temperatures rise, algae blooms often cloud the water of sinkholes or the surface becomes layered with a covering of duck weed. This tiny, free-floating plant often hides the clear water below. Many of the springs near rivers (Troy, Little River, Otter and others) are subject to flooding during periods of heavy rainfall. When the rivers tend to rise, their dark waters invade clear springs, making them unfit for diving. It is best to check with one of Florida's dive shops to learn the condition of the springs before planning your trip, especially in late summer and early fall.

Spring Area Activities

SCUBA DIVING. The larger springs and sink basins, such as Troy, Ginnie and Alexander Springs, provide the safest diving conditions in the world for certified scuba divers. Danger comes when those untrained in cavern or cave diving techniques venture into waterfilled caves beyond the natural glow of light. Because of this potentially deadly situation, commercial and state operated diving parks with spring caves do not allow untrained cave divers to carry underwater lights. This excellent rule prevents unwary divers from making the mistake of exploring dark inner passages.

SNORKELING. People without the proper training for scuba can still enjoy the underwater beauty of springs by using only a mask, fins and snorkel. Simply snorkeling around the surface of a boil or down its run can be a great adventure. Snorkeling is the only way to explore some of the springs where scuba is not allowed.

UNDERWATER PHOTOGRAPHY. Because of the superb water clarity in the springs, usually 100 feet-plus, they are ideal for underwater photography. The limestone cliffs, crevices and cave entrances make spectacular settings for either color or black and white shots. Most of the fish are accustomed to divers and can be approached easily. In many springs, a handful of bread will bring out bream by the hundreds. The white sandy bottoms reflect the sunlight, so there is little need for flash outside the caves. Care should be taken by the photographer not to stir up the sand or silt, though, as this will detract greatly from the picture.

RELIC HUNTING. Throughout the ages, the white sands of the springs and limestone bottoms of the clear rivers they feed have become a treasure chest of relics dating from prehistoric times through the development of man.

Mammoth and bison bones, giant shark's teeth, arrowheads, stone fish hooks, clay pipe bowls, old bottles and coins, and a vast assortment of other relics that trace Florida's development lie hidden, waiting to be uncovered by the patient collector. The beds of the Suwannee, Santa Fe and Chipola Rivers are often productive hunting grounds.

CAMPING. Camping is quite popular in the spring areas. Many of the commercially operated springs offer sites for truck campers and trailers with complete hookup facilities. Shower and toilet facilities can also be found. A few of the owners of springs located on private land allow camping, but only as long as their land and property are not abused. Several beautiful springs have been closed to the public during the past because of their misuse by campers and divers. We have heard story after story from these spring owners about the problems they encounter when they leave their areas open. The grounds and spring bottoms are littered with trash. Livestock has been chased and even killed. Game fish have been illegally speared. Remember, when you are using an area, the owner is bending to keep it open for you. End a dive by bringing up cans and bottles from the springs. Help protect your privilege to dive.

A few spring sites that cater to campers are: Vortex and Cypess Springs in the Panhandle; Ginnie and Otter Springs in the north central section; and Alexander, Juniper and Blue Springs in the State's center.

The springs offers caverns, camping and canoeing.   Bill Smith

TUBING AND CANOEING. Many of the large runs flow for miles throughout Florida's woodlands. Drifting in a slow current on crystal waters for miles and miles through green walls of vegetation is a peaceful experience, long remembered.

Florida is probably the only state where used tire tubes are at a premium. They are grabbed up on weekends and taken to some of the popular tubing runs in the State, where, with a little practice, you can quickly learn to maneuver the little rubber vessels with ease. Two of the most popular tubing runs are the Ichetucknee and Rainbow Rivers

Canoes are used for longer trips. Florida canoeing is a popular activity and the State has many excellent slow-water canoe trails. Divers have learned about the adventure of canoe trips down the Suwannee and Santa Fe Rivers. Such excursions can last for a few days or a few hours. While paddling down a river, springs can be easily located from where their runs empty into the river. Spend as long as you like exploring a spring—then continue downriver searching for new areas.

NATURE STUDY. A diver can spend many fascinating hours studying the underwater ecological systems of the springs. The clear water makes the springs one of the few places where freshwater habitats can be observed firsthand. The springs located near rivers have become the homes of many saltwater fish and mollusks as well as freshwater inhabitants, creating fascinating underwater communities. At Crystal River, it is common to see many varieties of salt and freshwater fish swimming around the boil area. In winter, manatees come into the bay and are a real attraction.

Springs and sinks are located in some of Florida's most beautiful woodland areas, so don't forget to save some time for topside exploration.

GLOSSARY OF SPRING TERMINOLOGY

Boil. Surface disturbance of the spring pool caused by the upward velocity of the spring's flow.

Chimney. A vertical tubular opening in the rock.

Crevice. A narrow crack or split in the rock.

M.G.D. Million gallons daily.

N.A.C.D. National Association for Cave Diving.

N.S.S.C.D.S. National Speleological Society Cave Diving Section.

Natural Bridge. Short section of ground covering surface of stream.

Permanent Line. Safety line permanently left in a cave.

Run. Water course flowing off from a spring.

Safety Line. White nylon line used to mark paths through caves.

FLORIDA CAVE DIVING

BY SHECK EXLEY

Since National Speleological Society (N.S.S.) divers first took scuba into a cave at Blue Hole Springs along the Ichetucknee River in 1951, cave diving has been pursued in Florida by hundreds of thousands of divers. Some enter the springs and sinks for serious speleological purposes such as exploration and survey, archaeology and paleontology, but most simply thrill to the warm, clear water—about 72 degrees year-round—and the fascinating limestone formations. Unfortunately, cave diving is not without its hazards: more than 150 divers have drowned while cave diving since 1960. While these victims represent only a small percentage of persons entering caves, N.S.S. recovery teams have found that, invariably, the accidents could have been avoided.

Why have these divers drowned? N.S.S. studies have shown that in every case, one or more of three major safety procedures were violated. Many perished because they became lost, having failed to run a single, continuous guideline from the entrance throughout the dive. Another major cause has been running out of air. At least two-thirds of the diver's air should be reserved for the trip out of the cave. Finally, the small percentage of fatal accidents where the victims did use a continuous guideline and had apparently planned their air reserves properly were all found to have involved diving to excessive depth. 130 feet has well been established as the maximum safe depth for sport divers, and this goes for cave divers as well.

In the past few years, most of the best cave diving spots have been closed to diving for various reasons. Some spots have been ruined by pollution caused by drainage wells, surface pollution, construction of canals and reservoirs, etc. Many springs that were clear a few years ago are now always murky. In some cases, health authorities have been forced to close spots because of contaminated water. A much larger percentage of these closed spots, however, have been restricted because of the

A safety line guides cavers through an underwater passage.

Pete Velde

thoughtlessness of some divers, from cutting down trees, carving names on walls and littering to making too much noise and riding a farmer's livestock. However, the biggest problem remains cave diving fatalities. Fear of lawsuits, bad publicity, and concern about accidents have forced many landowners to post diving areas. Fortunately, the number of cave diving fatalities has declined sharply since 1974, despite a probable increase in the number of participants. This is due in large part to the educational efforts of the N.S.S. Cave Diving Section and other volunteer diver training agencies.

Equipment and Procedures

Tanks should not have "J" valves. These are easily knocked down and have been known to malfunction. A dual-valve manifold should be used in conjunction with two independent, single-hose regulators so that if one regulator should malfunction, you can simply cut off the air to the bad regulator, then switch to the other one and make a safe exit from the cave. One of the regulators should be equipped with a longer hose to facilitate buddy-breathing, as well as a submersible air pressure gauge. This gauge is all-important: the entire dive revolves around the diver's air supply. Ever since it was founded in 1941, the N.S.S. has advocated the use of three independent sources of light for each caver, and this is just as important for cave divers. A buoyancy compensator, either front- or back-mounted, is necessary to provide lift to stay off the silt on the bottom. Watches, depth gauges and submersible tables are needed to monitor decompression requirements. Finally, the cave diving team must have a good nylon safety line of at least 160 lbs. tensile strength on a good underwater reel. Of course, all open-water gear, such as tanks, wetsuits, masks, fins and knives are also used, with the exception of a snorkel.

A dive plan must be formulated before entering the water, including planned depth, penetration, bottom time and air turnaround. N.S.S. recommends the "third rule," where the cave diving team automatically turns around and starts out of the cave as soon as the first member exhausts one-third of his original air supply. The feeling is that two-thirds or more left would be sufficient to cope with any emergency situation. One diver is designated team leader. This diver will be the first one into the cave and the last one out. The team leader secures the line once just outside the cave entrance underwater; then once just inside, before daylight is lost. As the leader deploys the safely line from his reel, his teammates follow him closely, maintaining contact with the line at all times. The divers also maintain contact with each other by glancing back, watching the reflection of each other's lights, hand signals, etc. If the desired turnaround is reached, or the dive is called by any member of the team, then the entire team starts out of the cave together. It is very important that any member of the team feel free to call the dive for any reason at any point. Not infrequently, experienced cave divers abort dives for no reason other than simply "feeling a little uptight." Throughout the entire dive, the team stays up near the ceiling to conserve air and to keep from stirring up the silt on the bottom.

Types of Caves

There are four general types of underwater caves in Florida—springs, sinks, spring-siphons and natural bridges. A spring flows out of the ground, forming a run that goes to a nearby river, ocean, etc. A sink has no discernible flow and is generally enclosed by a sunken area known as a depression. A spring-siphon has two caves, one with water coming out, the other with water going into the ground. When the river rises in periods of heavy rainfall, causing dark water to "back-up" over the springs, many springs reverse their flow to become siphons, draining the dirty river water. Natural bridges cover surface streams, such as that over the Santa Fe River at O'Leno State Park. Most of these surface streams have poor visibility. However, some springs have short, diveable natural bridges over their runs.

Hazards

Cave diving involves surmounting a number of natural hazards, the most important of which is the cave ceiling. Should any emergency occur while cave diving, the distressed diver cannot make a free ascent to the surface. He must rely on his knowledge, skill, equipment and experience—and that of his partner as well. Other important hazards include currents, visibility, and various physical peculiarities of the cave. Current can lead to overexertion, and certainly make buddy-breathing much more difficult. Siphons are to be avoided—the strong downward flow makes it necessary to use much more effort and air to swim out of the cave, which could possibly be critical should a problem arise. In periods of warm weather, many sinks and springs become cloudy with algae growth. Heaviest growth (and poorest visibility) is usually in the surface layers. Most caves also contain deposits of silt on walls, ledges and especially the bottom. A careless movement, such as a flipper stroke, can cause the soft sediment to stir up instantly—reducing visibility from hundreds of feet to inches. Often, the diver is not aware of this problem until he glances behind.

Many caves branch off into underground mazes, making a safety line mandatory to find one's way out. Restrictions (narrow areas in caves) are to be avoided. Deep caves should also be avoided because of potential problems with narcosis, decompression, etc. Perhaps the best rule to follow in deciding whether or not to penetrate a given cave is to ask oneself the following question: "If anything were to happen in the cave at the worst moment, could I still be certain that I could get my buddy out alive? Or he get me out?" If the answer is no, only a fool goes on.

Emergency situations are unlikely to arise for the well-prepared cave diving team. However, everyone entering a cave should be prepared to deal with them. A few of these emergency situations include: air supply loss, lost diver, line entanglement, broken safety line, loss of visibility (light failure or "silt-out"), hurt diver and unconscious diver. The procedures for coping with these situations should be mastered with your buddy under simulated conditions in open water at night before entering an underwater cave, and reviewed together frequently.

It is well to remember that a cave diving accident may not result in one diver's death alone. Many cave diving accidents are multiple drownings—the initial victim's partner may well die also. If he drowns in a hazardous area, he may jeopardize the lives of a body recovery team. His drowning may also result in not only the closing of that particular cave to all diving, but also other caves and, possibly, an entire county. However, worst of all—at least from the victim's standpoint—is the tragic effect of grief that the accident will have on his immediate family and friends. And for what? A foolish dive that could easily have been made safely if the diver had been willing to wait and learn cave diving the safe way.

The Safe Way

The minimal training received in a basic scuba course is simply not enough for cave diving. A majority of victims have been certified basic scuba divers, but, nevertheless, drowned before making their fifth cave dive. Fortunately, the Cave Diving Section of the N.S.S. now offers an inexpensive introduction to cave diving through its two-day "Cavern Diver" certification program. This qualifies divers to "ledge dives" in the shallow areas just inside large cave entrances where there is plenty of natural light. Persons interested in more serious cave diving may enroll in the more extensive N.S.S. "Cave Diver" certification course, which usually takes eight days or longer to complete. Other national agencies offering this training include NAUI, PADI and YMCA.

The N.S.S. Cave Diving Section also offers periodic cave diving workshops all over the country, where novices and experts alike can get together and exchange ideas on various aspects of cave diving. Such workshops usually include actual guided practice dives as well as lectures and discussions in an informal atmosphere. Typically, the workshops are held in Branford, Florida, each year on Memorial Day weekend and New Year's weekend, as well as summers at the N.S.S. Convention.

Ned DeLoach

Publications

In addition to cave diving courses and workshops, the N.S.S. Cave Diving Section has made available several current publications regarding cave diving. "Cave Diving Safety" is a free pamphlet published as a public service by the Section that briefly describes many of the basic procedures required for cave diving. Safety maps of the most popular springs in Florida are available in large blueprint format at $3.00 each, as is a 25 cent brochure explaining their use. Basic Cave Diving: Blue Print for Survival ($2.95) by Sheck Exley gives an introduction to the fundamentals of cave diving in Florida. The N.S.S. Cave Diving Manual ($11.95) is America's first complete authoritative text on the subject, with more than 300 pages covering virtually every aspect of cave diving. For information on any of these books, courses or workshops, write the N.S.S., CDS Publications, H.V. Grey, P.O. Box 575, Venice, Florida 34284-0575.

Conservation

You can choose whether or not you want to risk destruction by foolishly taking unnecessary risks while cave diving, but the cave has not such choice. It is there, and it is vulnerable to vandalism. Caves have unique scientific, recreational and scenic values that should be preserved for future generations to study and enjoy. N.S.S. members must pledge to do nothing that will deface, mar or otherwise spoil the natural beauty and life forms in caves. Be kind to the cave, and remember the N.S.S. motto:

"Take nothing but pictures

...Leave nothing but footprints

...Kill nothing but time."

The N.S.S. Cave Diving Section

The N.S.S. Cave Diving Section is by far the largest cave diving organization in the western hemisphere. Its members, novice and expert alike, engage in a large number of activities to encourage safe, productive and enjoyable cave and cavern diving.

Membership in the N.S.S. is open to all persons interested in caves. Applications may be obtained by writing to the National Speleological Society, Cave Avenue, Huntsville, Alabama 35810. Any N.S.S. member may join the Cave Diving Section by sending $5.00 to Sandy Fehring, 3508 Hollow Oak Place, Brandon, Florida. Non-members may subscribe to the bimonthly magazine of the Section, Underwater Speleology, by sending $8.00/year to Sandy Fehring.

Biographical Note

Sheck Exley is the most experienced cave diver in the world, with more than 3,000 cave dives logged in many countries on four continents. He is a charter member of the N.S.S. CDS and was its first Chairman. For his cave rescue work, he is the only diver ever to receive the Distinguished Service Award of the Florida Sheriffs Association. He has also received the Lew Bicking Award as America's top cave explorer in 1981, the Abe Davis Cave Diving Safety Award, and is a Fellow of the N.S.S. and the Explorers Club. He has written more than 100 articles and six books on the subject of cave diving safety.

1

BRANFORD

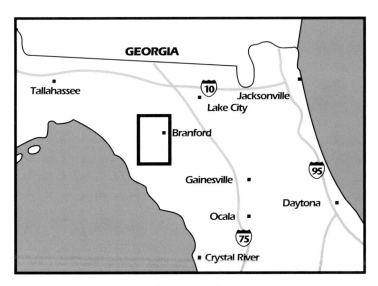

Spring Diving Locations

1. Charles Spring
2. Blue Spring-Lafayette Co. (Snake, Yana, Bob Cat & Stephens Sinks)
3. Telford Spring
   Telford Sink
   Terrapin Sink
4. Bonnet Spring
5. Peacock Springs
   Pot Hole Sink
6. Olson Sink
7. Challenge Sink
8. Orange Grove Sink
   Cistern Sink
9. Running Springs
   Cow Spring
10. Convict Spring-River Rendezvous
11. Royal Spring
12. Troy Spring
13. Little River Spring
14. Branford Spring
15. Rock Bluff Spring
16. Sun Spring
17. Hart Spring
18. Otter Springs
19. Fanning Spring

NOT ON MAP
   Suwannee Spring
   Manatee Springs
   Blue Springs-Madison Co.
   Blue Sink-Madison Co.

Canoe Trail with Spring Sites
Suwannee River

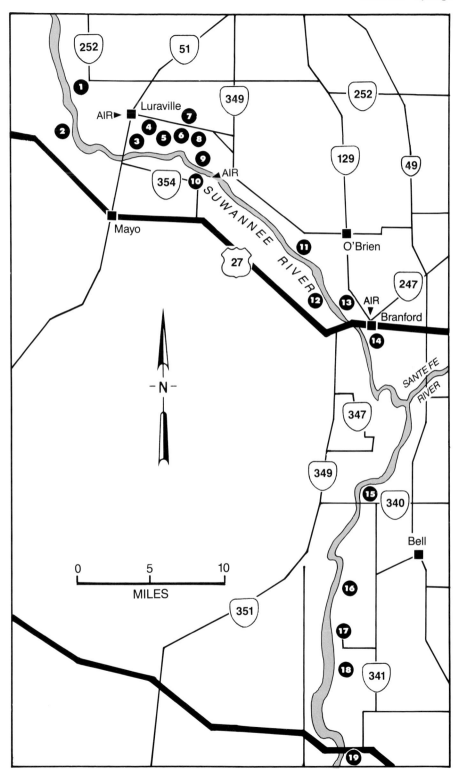

Area Information

Branford, located on the east bank of the famous Suwannee River, has long been the center of spring diving in Florida. Dozens of springs are located both north and south of the small town. The Peacock Slough area which became a State Park in 1990 is one of the most popular groups of springs in the state. A diver can enjoy a dozen different springs and sinks within a one-and-a-half-mile radius. Troy Springs, located on the west bank of the Suwannee, has a large open basin with a depth to 80 feet. The interesting remains of an old paddle-wheeler lie in the middle of the run near the river. A motel, restaurants and a complete dive shop are located in Branford.

There are air stations at River Rendezvous, fifteen miles northwest of Branford on the west bank of the Suwannee River, and on the road between Luraville and the Peacock Springs State Park entrance.

NO.1 CHARLES SPRING

Location: From the bridge over the Suwannee River on S-51, travel east 5.5 miles. Turn left on S-252 and go 3.1 miles to sharp right bend in road. Go straight at the bend on small dirt road 1.4 miles to the springs.

The head pool is about 50 feet in diameter, with a 10-foot depth to cave entrance. There are two natural bridges in the run that can be free-dived. The cave is dangerous due to extreme silt and cave-ins. There is an open area near the spring for camping, with a boat ramp to the Suwannee.

NO.2 BLUE SPRING-LAFAYETTE CO.
(SNAKE, YANA, BOB CAT AND STEPHENS SINKS)

Location: From Mayo travel northwest on US 27 for 5 miles to S-251-B. Turn right and go 2.2 miles. Turn right (east) onto a sand road just before a guardrail. In .8 mile, turn right to Yana Sink or continue .1 mile further to Blue Spring.

In 1985, Lafayette Co. developed the Blue Spring area into a park withrecreational facilities and a boat ramp. There isan entrance fee. Scuba is not allowed in Blue Spring but the sinks are open for diving.

Blue Spring, situated on the west bank of the Suwannee River, is a 25-foot basin skirted by a high bluff. About halfway down the short run leading to the river is a large, natural bridge.

A large cave entrance opens into a wide room before narrowing again into a 300-foot tunnel that leads to Snake Sink at a depth of 47 feet. The water of the spring and connected sinks has a green tint and an average visibility of 50 feet. Snake Pool is about 250 feet long and 30 feet wide. A cave entrance at the west end runs for 20 feet at a 45-foot depth to Yana Sink.

Yana's basin is 150 feet long and 70 feet wide and has a maximum depth of 35 feet. The cave at the west end leads into the Dome Room. The larger room's most prominent features are two large solution domes. The first is 40 feet across and has an air pocket where divers can converse. Only those with advanced cave diving training should proceed beyond the Dome Room. Stop when depths drop to 60 feet. The system continues for over 20,000 feet back into the woods. Bob Cat, Stephens I, II and III, and Skiles Sinks all open the system to the surface at various points along the channel.

NO.3 TELFORD SPRING

Location: Travel north on S-51 from Mayo toward Luraville. After crossing the bridge over the Suwannee, turn right on the first road (dirt road across from truck inspection station). Go a short distance to the stop sign and turn right. Continue .9 mile, bearing to the right as the road continues straight to a boat ramp.

Telford Spring is located on the eastern shore of the Suwannee River. Bare dirt banks slope gently down to the small basin, which feeds a run about 40 yards long. A small crevice ends in a small cave.

NO.3 TELFORD SINK

Location: Just north of dirt road before reaching Telford Spring, about 200 feet from the spring.

This small sink has three round limestone shafts which connect underwater in a room with depths to 33 feet. Two small, extremely silty tunnels lead off from the room.

NO.3 TERRAPIN SINK

Location: In a shallow depression, just past Telford Spring, about 150 feet west of dirt road. A footpath leads to the sink.

This small sink is believed to be a spring-siphon that connects to Telford Sink. Depths are to 35 feet.

NO.4 BONNET SPRING

Location: Go north on S-51 from Mayo. After crossing the bridge over the Suwannee River, go 1.7 miles. Turn right on paved road across from the service station/store. Go 2 miles and turn right at the first dirt road past the fence line. A gate now blocks access.

Bonnet Spring is a small spring in a beautiful setting. There is interesting snorkeling in rock crevices and through a large area of lily pads filled with fish.

NO.5 PEACOCK SPRINGS

Location: From Mayo going north on S-51, travel 1.7 miles from the middle of the bridge over the Suwannee. Turn right on paved road across the highway from the service station/store in Luraville. (This is the second road on the right after leaving the bridge.) Travel 2.2 miles and turn right at the entrance to the park. Continue straight for .5 mile to parking area.

OR

From Branford travel north for 6 miles on US 129 to O'Brien. Turn left on S-349 and go 12 miles. Turn left on another paved road (watch for two green dumpsters) and go 1.5 miles and turn left again at stop sign. Continue for 3.5 miles. Turn left at the park's entrance.

This area, known for years as Peacock Slough, includes several diveable springs and sinks. There are a few open water areas but most diving is in water-filled cavern rooms or down dark passages that wind for over 21,000 feet under the thick tangle of woods above.

After many years of changing ownership this popular dive site became a State Park in 1990. The park opens at 8 AM and closes at sunset. No camping or alcohol is allowed. Divers must show C-cards. Underwater lights are not permitted unless a diver has cavern or cave certification. There are a few picnic tables and two

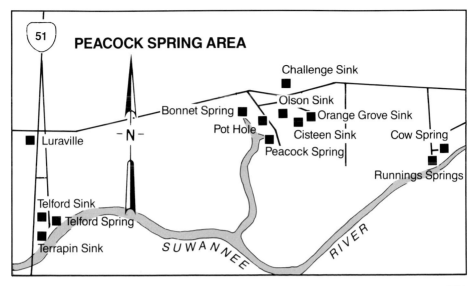

51 **PEACOCK SPRING AREA**

Challenge Sink

Bonnet Spring

Olson Sink

Orange Grove Sink

Pot Hole

Cisteen Sink

Cow Spring

- N -

Luraville

Peacock Spring

Runnings Springs

Telford Sink

Telford Spring

Terrapin Sink

SUWANNEE RIVER

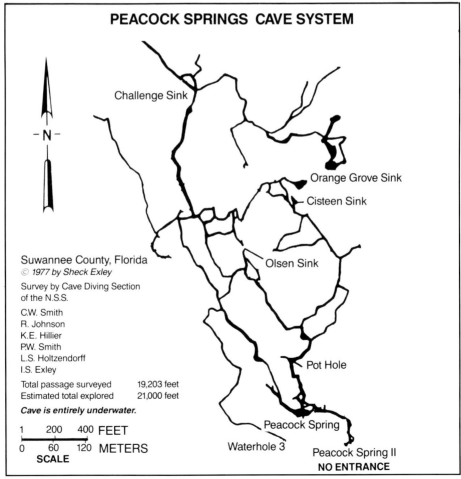

PEACOCK SPRINGS CAVE SYSTEM

- N -

Challenge Sink

Orange Grove Sink

Cisteen Sink

Suwannee County, Florida
© 1977 by Sheck Exley

Survey by Cave Diving Section
of the N.S.S.

C.W. Smith
R. Johnson
K.E. Hillier
P.W. Smith
L.S. Holtzendorff
I.S. Exley

Olsen Sink

| Total passage surveyed | 19,203 feet |
| Estimated total explored | 21,000 feet |

Cave is entirely underwater.

Pot Hole

1   200   400 FEET
0   60   120 METERS
SCALE

Peacock Spring

Waterhole 3

Peacock Spring II
NO ENTRANCE

portable outhouses. A small fee must be paid for each vehicle that enters the Park but the diving is free.

Three beautiful springs feed a 1.5 miles slough that runs to the Suwannee. The headspring offers a good cave dive, with a tunnel that winds its way to Pot Hole Sink, about 75 yards from Peacock. The mouth of the cavern starts in about 18 feet of water and is about nine-by-five feet, making access to the first room easy. The cavern room is large and extends east to west. At the mouth is a permanent line that runs to the west end of the room, where it ends at the ceiling about 15 feet off the floor. At the north end of the room, at a depth of 45 feet, is a sign that warns those who are not trained cave divers to venture no farther.

Those divers with proper cave training and experience may want to continue on to Pot Hole Sink. Below the sign, a three-by-ten foot slit drops to a tunnel with depths of 65 feet. The winding tunnel, with several off-shoots, leads about 400 feet to Pot Hole Sink. Surface light is difficult to spot at Pot Hole. Proceed up the narrow shaft one at a time.

NO.5 POT HOLE SINK
Location: About 75 yards north of Peacock Slough. It is 20 feet to the side of the entrance road.

Pot Hole Sink's small basin drops 30 feet to water level. Caution should be exercised in entering the sink. The water at the surface is six feet by ten feet, but four feet under it narrows to a small three-foot chimney. A submerged tree limb and loose sand increase the hazards of entry. The narrow passage spirals down

15 feet, and begins to widen gradually as you approach the main tunnel from Peacock.

From Pot Hole there is also a connecting tunnel that travels 1,200 feet at a depth of 50 to 70 feet to Olson Sink. This, as well as other cave transverses, should be undertaken only by certified, experienced cave divers.

NO.6 OLSON SINK

Location: The Sink is inside the Peacock Springs State Park. Turn left off the entrance road .3 mile from the gate. (Look for small Orange Grove sign.) Go .1 mile; the depression on the right is Olson.

This is a beautiful sink with steep walls located deep within the woods. The dive is for certified cave divers only. Take care carrying equipment to water level. The narrow path is quite slick when wet.

On the north side, at a depth of 20 feet, a tunnel opens into a small room at a depth of 65 feet. A silty passage winds its way to Challenge Sink. The entrance to the south tunnel is partially hidden by fallen limbs. This tunnel runs to Peacock Spring at a depth of 70 feet.

NO.7 CHALLENGE SINK

Location: Directly across the paved road from the old turn-off to Olson(The access from the paved road to Olson is now blocked off, as is the footpath that leads to Challenge.)

This is a small sink, 10 feet deep, with a narrow entrance on the south side that opens up and drops to a depth of 70 feet, where it leads to Olson Sink. This is considered an advanced cave dive.

NO.8 ORANGE GROVE SINK

Location: Inside the Peacock Springs State Park. Continue past Olson Sink for .2 mile.

This is a large sink located in a rugged setting. Massive limestone cliffs drop to the surface water, which is covered through much of the year with a thin green layer of duckweed. When local water table conditions are right, the visibility can exceed 100 feet. The bottom slopes to a depth of 60 feet, where it is crisscrossed with large tree trunks and limbs. The cave is located on the north wall of the sink at a depth of 53 feet. The entrance measures 15 by 7 feet, and opens into a large room. This room is about 60 by 80 feet. The depth is 65 feet at the ceiling and 100 feet, where large boulders cover the floor.For the experienced cave diver only, there is a five-by-eight-foot entrance to a corridor at the 70-foot level. This tunnel winds for 75 feet before entering the Throne Room. Conditions here are dangerous due to a deep layer of fine black silt.

NO.8 CISTERN SINK

Location: The sink's depression can be seen off to the right from the bluff overlooking Orange Grove Sink.

From the surface, this sink resembles nearby Orange Grove. The diameter is about 30 feet, depending on the height of the Suwannee. There is a cave entrance in 10 feet of water that leads to a corridor which winds down 50 feet to a room called the Witches' Kettle. The room is named for its precarious entrance and extremely silty condition. From here, the corridor winds on and on. This dive should only be made by trained, experienced cave divers.

NO.9 RUNNING SPRINGS

Location: Travel north on S-51 from Mayo toward Luraville. Go 1.7 miles past the center of the bridge over the Suwannee. Turn right on the paved road across the highway from service station/store (same as the road to Peacock). Travel 3.8 miles. Turn right on the dirt road across from a small church and go .8 mile to where the main road bears to the left, but continue straight ahead on the small dirt road .6 mile to the springs.

Two beautiful springs are located on a high bluff overlooking the Suwannee. These springs are not large or deep enough for scuba but are a joy to snorkel or swim. The right spring is 10 feet deep with a short run that siphons under the river bank to the Suwannee. The left spring has an underwater natural bridge with a 3-by-5-foot opening that can easily be free-dived.

NO.9 COW SPRING

Location: Follow the directions to Running Springs, but bear left on the small dirt road .2 mile before arriving at Running Springs.

Cow is a small-siphon spring filled with crystal clear water. The cave is beautiful to dive but limited in size. There are two entrances leading into a small room that goes back about 25 feet. Maximum depth is 40 feet. There are two tunnels leading from the room. The tunnel on the left is the spring entrance. It is restricted in size. The siphon entrance, at the back of the room, is also narrow and a difficult dive.

NO.10 CONVICT SPRING—RIVER RENDEZVOUS

Location: From Branford travel NW on US 27 for 12 miles. Turn right on C-354 Convict Spring Road. (Watch for a large red and white sign across the highway from a small store.) Go 1.6 miles on this paved road until it makes a sharp left. Continue straight on a dirt road for .6 mile. Check in at the office.

This is a commercially operated diving park. It was opened in 1986 on the land around Convict Spring.

The large wooded site is on the west bank of the picturesque Suwannee River. Care was taken to provide complete facilities for the visiting diver. Camping grounds and completely furnished rooms are available for the day or week. A dive shop with a 5000 psi compressor and a restaurant are on the grounds. A shuttle bus makes regular pick up and delivery runs to the Peacock spring system, as well as other nearby diving locations. Canoe rentals are available.

Convict Spring is located directly in front of the lodge. The clear water basin is about 50 feet across with depths of 25 feet. A cave can be penetrated for 80 feet to a maximum depth of 30 feet. A small run winds to the Suwannee.

NO.11 ROYAL SPRING

Location: From Branford, travel north for 6 miles on US 129 to O'Brien. Turn left on S-349 and go 9 miles. (Just after highway bears right watch for two green dumpsters on the left.) Turn left on graded road and go .7 mile; then turn left on sand road (This turn is just before the white trailer house.) Continue .2 mile to the spring.

JIM HOLLIS' RIVER RENDEZVOUS

ROUTE 2 - BOX 60 • • • MAYO, FL 32066

(904) 294-2510 • Fax 1-904-294-1133 • 1-800-533-5276

• Nestled on the beautiful Suwannee River •
• A complete Full-Line Pro-Dive Shop with 5000 PSI Compressor & Haskel •
• Motel-Lodge Facilities with Restaurant & Suwannee River International Beer Parlour •
• Featuring over 300 different kinds of Beer •
• Over 30 different Wine Coolers •

• Campsites Available with Hot Showers •

• RV Hook-Ups Available •

• DAY TRIPS AND NIGHT DIVES AVAILABLE TO TROY SPRINGS DAILY •

A-Frame & Mobile Home Trailers also Available on a Nightly of Weekly Basis.

OUR FACILITIES ALSO INCLUDE:

- 2 Hot Tubs
- Sauna
- Fishing
- Trampoline

- Steam Room
- Canoe Rentals
- General Store
- Game Room w/Pool Tables
- Three Day Private Scuba Classes Available•

- Tube Rentals
- Volleyball
- Pistol & Rifle Range
- 8' x 30' Boat & Swimming Dock

EVERYTHING IS BUILT AROUND OUR VERY OWN CONVICT SPRINGS

Don't Miss this Very Special Place • Jim Hollis' River Rendezvous,

Home of the Weekend Scuba Class.

Air Available until 9:00 P.M. - 7 Days a Week

• DIVE THE LOWER ICHETUCKNEE RIVER - TRIPS NOW AVAILABLE
• DIVE TROY SPRINGS IN COMFORT OFF OUR 24' PONTOON BOAT

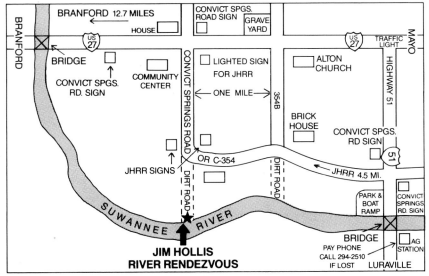

"ONCE IS NOT ENOUGH"

This large spring surrounded by steep banks is murky much of the year. When clear the water retains a greenish tint. A concrete retaining wall is built on the east side with steps to the water. The basin is about 40 yards in diameter, with a shallow 60-yard run to the river. A large cave entrance is located about 50 feet down the side of the limestone cliff. There is not much flow, and the cave is very silty and dangerous. Stay out! To the left of the run is a good boat ramp. This area is now a county park.

Troy Springs paddle-wheeler wreck.   Ned DeLoach

NO.12 TROY SPRING

Location: From the intersection of US 129 and US 27 in Branford, travel northwest on US 27 for 5.3 miles. Turn right on the paved road (across highway from white house) and go north 1.3 miles. As the paved road curves left, you will see a small green house trailer on the right. Turn onto the dirt road that runs between the trailer and a fence line for .6 mile to the spring. Editor's note: The access to Troy was closed in 1989 after the property changed ownership. Hopefully it will be reopened in the near future. In the meantime access can be make by river. This is legal as long as you remain in the basin or its run.

Troy is a large spring located on the west bank of the Suwannee River with depths to 80 feet. Because of its great size, depth, clear water and lack of a cave, it is one of the State's primier locations for check-out dives.

The wide run has depths from four to six feet. At the end of the run near the river lie the broken remains of the old Suwannee River steamboat *Madison*. She was scuttled in September 1863 by her captain, who abandoned his vessel to lead a company of Confederate soldiers in Virginia.

NO.13 LITTLE RIVER SPRING (Restricted to Certified Cave Divers)

Location: From the intersection of US 27 and US 129 in Branford, travel north on US 129 for 3.1 miles. Turn left on paved road at large Camp O'the Suwannee sign. After 1.8 miles, bear left at the second sign.

Little River is rated among Florida's finest cave dives. The spring basin is large, with a bare limestone and sand bottom. There are few aquatic plants or fish. A strong flow issues from a small cave, 10 feet in diameter, in about 15 feet of water. The cave slopes to a level of 60 feet; then it makes a sharp left turn where all surface light disappears. At this point, all inexperienced cave divers should end their penetration. There is a permanent sign which serves as a reminder.

Continuing past the sign, the experienced cave diver enters a complex and silty maze that leads to a small room with a chimney that drops from 70 to 100 feet. At 100 feet, the cave levels off into a picturesque winding tunnel known as the "serpentine passage." The passage leads to a large section, 800 feet from the entrance, that has been named the "Florida Room." Past the big section, the cave begins to branch, gradually becoming smaller and more silty.

NO.14 BRANFORD SPRING

Location: 100 feet from the southeast corner of the bridge over the Suwannee in the town of Branford, adjacent to the Branford Dive Center.

This is a great place to cool off or check out your gear. The clear, blue spring is about 65 feet across, with depths to 15 feet. A wooden deck and ladder allow easy access to the water.

NO.15 ROCK BLUFF SPRING

Location: From the intersection of US 27 and US 129, 4 miles east of Branford, turn right on US 129 and travel south for 14 miles. Turn right on C-340, traveling west to the white inspection area next to a boat ramp. Launch your canoe or boat here, and go upstream (north) for about 150 yards. Watch for the clear spring run on the right side of the river.

A scenic, cypress-lined run winds inland for 40 yards to a large, shallow basin. There are many bream, mullet and turtles in the clear, inviting water. The main dive is into a rock fissure, 60 feet long and 20 feet wide. The edge is in 10 feet of water and drops to a small cave entrance at 35 feet.

The cave system is extensive but narrow, difficult and dangerous. A strong flow issues from the small cave. All the surrounding land is privately owned. A home is next to the spring. Absolutely no trespassing is allowed. It is all right to explore the spring as long as you stay in the water. Anchor in the basin. Don't pull your boat ashore. Please be quiet and leave no litter.

NO.16 SUN SPRING

Location: From Branford travel 4 miles east on US 27. At the blinking caution light turn right on US 129 and go south for 14 miles. Turn right and travel west on C-341 for 3 miles before C-341 makes a sharp left turn. Turn left and go south for 1.7 miles. Watch for the Sun Spring sign. Turn right (west) and follow the road for 1.5 miles. Turn left and go one-quarter mile to the spring area.

The Sun Spring area is being developed for housing. Large groups are not encouraged to dive here. A canal from the Suwannee has been cut to the spring for waterfront property.

The spring is small but interesting. A circular rock cliff 10 feet below the surface drops to a depth of 20 feet. This is a nice place to snorkel with plenty of bream and turtles always about.

Running Springs on the bank of the Suwannee.   Ned DeLoach.

NO.17 HART SPRING

Location: Follow the directions from Branford toward Sun Spring. Where C-341 makes a sharp left (south) go 4.1 miles. Turn right (west) and follow the road for about 2 miles to the entrance of Gilchrist County Recreational Park.

Hart Spring is a pleasant family recreation park with picnic tables, refreshments and a large swimming area. Diving is not permitted in the boil. A large run lined with cypress trees flows a short distance to the Suwannee.

NO.18 OTTER SPRINGS

Location: Follow the directions from Branford toward Sun Spring. Where C-341 makes a sharp left (south) go 6.7 miles. Turn right and go 1 mile before bearing right onto paved road that leads to the park entrance where you must register.

OR

From town of Fanning Springs on US 19-98 take S-26 north to town of Wilcox; take C-232 1.7 miles. (Look for Otter Springs RV Park sign.) Turn left and go 1 mile before bearing right.

Otter Springs is a complete RV resort with many organized activities. A dive shop was opened in 1990 on the edge of the spring basin. A fee for diving is charged. Cave and cavern dive courses are taught. Canoe rentals are available.

The headspring is located in a small basin with a long, narrow crevice which drops down about 25 feet to a small cave. The second crevice runs parallel to the first, with depths to 14 feet. A small, circular tunnel connects the two, and a small corridor runs from the cave. It is silty and has a weak flow. Snorkeling is good over the basin and down the run. There are excellent camping facilities with trailer connections.

NO.19 FANNING SPRING

Location: In the small town of Fanning Springs on the east bank of the Suwannee River. Turn right off US 19-98 on the second road past the bridge.

A large, open basin on the east side of the Suwannee River, The spring pool is 30 yards across with a maximum depth of 20 feet. The water issues from a sand boil.There is no cave. The land is owned by a church group and there is a 50 cent admission charge. There is a dock, concession stand and boathouse. No camping is permitted.

SUWANNEE SPRING (Not on Map)

Location: From Live Oak, travel north on US 129 for 8 miles to the Suwannee River Bridge. Turn east on the south end of the bridge and go .1 mile to the spring.

An interesting rock formation in the spring issues about 20 m.g.d. of clear green sulfur water. The entire spring is surrounded by rock walls that look like an old Spanish fort.

MANATEE SPRINGS (Not on Map)

Location: Just north of Chiefland on US 19 (watch for green sign) turn west on S-320 and follow signs about 6 miles to Manatee State Park entrance.

Guests planning to dive at Manatee must register and pay the required fee when entering the park. A C-card is required. Cave divers must show cave

certification. Divers without cave training are not allowed to carry lights underwater. This rule is strictly enforced. There are two good dives here: Manatee Springs and Catfish Sink. Both are a 100 yard hike from the parking area. It is best to park at the far (west) end. From here Catfish is straight ahead and to the right. The 120-foot circular basin is often covered with a green layer of duckweed. Entry is down a gentle slope to water level. A shallow limestone shelf extends 10 to 15 feet into the basin. The basin is about 35 feet deep and will easily silt up with much diving activity.

Manatee is a large spring basin that discharges 96 m.g.d. into a beautiful cypress-lined run that winds for almost a quarter mile to the Suwannee. A limestone cliff drops to a cave at a depth of 40 feet. A strong flow issues from the cave; it is very silty. A small wooden deck provides an easy entry. The basin and wide run are filled with aquatic plants and fish. Diving and swimming is restricted to the basin and the first 200 feet of the run. Water temperature is a constant 72 degrees.

Camping and picnic facilities are excellent. There are dozens of tables and grills between the sink and spring, and two covered picnic areas. A well-run concession has a good selection of food and also offers canoe and tube rentals.

A raised boardwalk follows the run's south bank for a quarter mile to the Suwannee. The deck extends over the river for a short distance and provides a picturesque view. This beautiful cypress-lined walk is not to be missed.

Although the park and spring receive their name from the famed sea cows, only a few now venture up the run to escape the cool river water in the winter. Hopefully they will begin to return now that boat traffic is not allowed in the run.

BLUE SPRINGS-MADISON CO. (Not on Map)

Location: From Madison, go east on US 90 for 2.5 miles. Turn left on S-6 and travel about 6 miles to the bridge over the Withlacoochee River. Turn right (south) on the west side of the bridge and go .1 mile to the spring.

Blue at Madison is a large spring, about 25 yards in diameter, and is situated in a basin with 30 feet of sand and rock cliffs. A near horizontal cavern opens to 20 feet by 30 feet at a depth of 30 feet. The strong flow is rated 94 m.g.d. The short run to the river is lined with cypress trees. There is a large area for camping with no facilities.

BLUE SINK-MADISON CO. (Not on Map)

Location: From S-6 over Withlacoochee River, go west on S-6 for .3 mile to graded dirt road on right. Go .5 mile to second sharp bend to the left; continue straight ahead onto another dirt road and go about 1 mile to the parking area 200 feet west of the sink.

This is a large sink, about 300 feet by 100 feet, on the edge of a swampy area. Visibility averages 50 feet, with depths to 45 feet on a rock crater near the middle. Some fish and turtles are present.

Cow Springs   Ned DeLoach

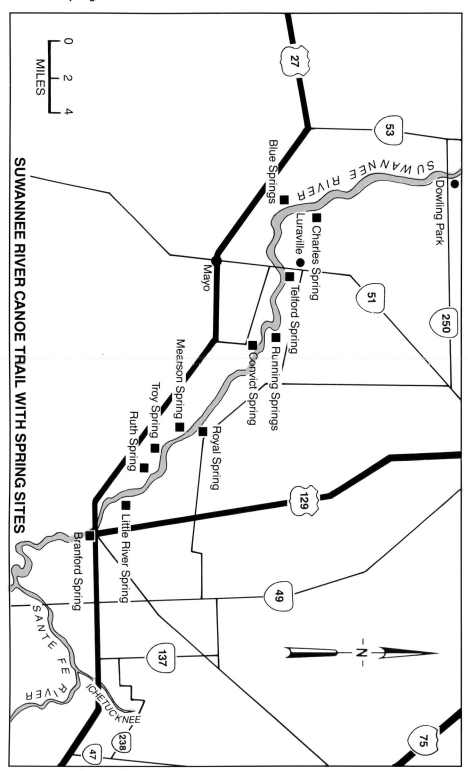

SUWANNEE RIVER CANOE TRAIL WITH SPRING SITES

The Suwannee River twists and turns for over two hundred miles from Fargo, Georgia to the Gulf of Mexico. South Georgia's Okefenokee Swamp gives birth to the famous river, but over twenty springs add to its flow before it empties into the sea. Many of these springs are dive locations mentioned in this book.

An enjoyable way to visit these springs is by canoe down the Suwannee. Bill Smith, located at the Steamboat Motel in Branford, has for years been carrying canoeists upriver for a pleasurable canoeing-diving trip back to Branford. He suggests two different length trips, depending on the amount of time you care to spend on the river. The two-day trip starts at Dowling Park 37 miles upriver from Branford. The night can be spent camping at Running Springs (No.9). This splits the trip in half with approximately 18 miles of river covered each day. For a one-day trip of seven hours you can begin at Royal Spring (No.11). This is 14 river-miles above Branford.

Below is a list of springs on or near the river that you can visit during your trip downstream. Their approximate distance from Branford is given to the left. The (No.) is a reference to their description in this chapter.

37	Dowling Park (town)
30	Charles Spring (No.1)
27	Blue Spring-Lafayette Co. (No.2)
22	S-51 bridge near Luraville (town)
21	Telford Spring (No.3)
18.5	Running Springs (No.9)
	3 mile below Running is an old railroad bridge
16.5	Convict Spring-River Rendezvous
14.5	Royal Spring (No.11)
10.5	Boat ramp
10	Mearson Springs
	on the west bank, up a wide run 75 feet in length; private property, not listed as a dive site
6.5	Troy Spring (No.12)
4.8	Ruth Spring
	on the west bank, a long narrow run leads to the spring, private property, not listed as a dive site
4.2	Little River Spring (No.13)
0	US 27 bridge, Branford, Branford Spring (No.14)

2

FORT WHITE – HIGH SPRINGS

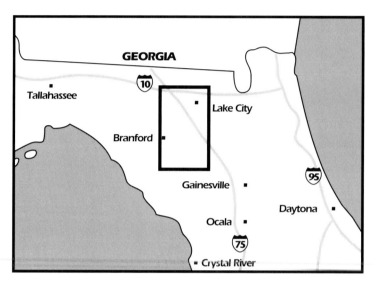

Spring Diving Locations

1. Ichetucknee Springs
2. Blue Hole Spring
3. Ginnie Springs
   Devil's Eye Spring
   Dogwood Spring
4. July Spring
5. Blue Springs-Santa Fe River
   Naked Spring
6. Rum Island Spring
7. Lily Spring
8. Poe Spring

NOT ON MAP
   Blue Grotto

Canoe Trail with Spring Sites
Sante Fe River

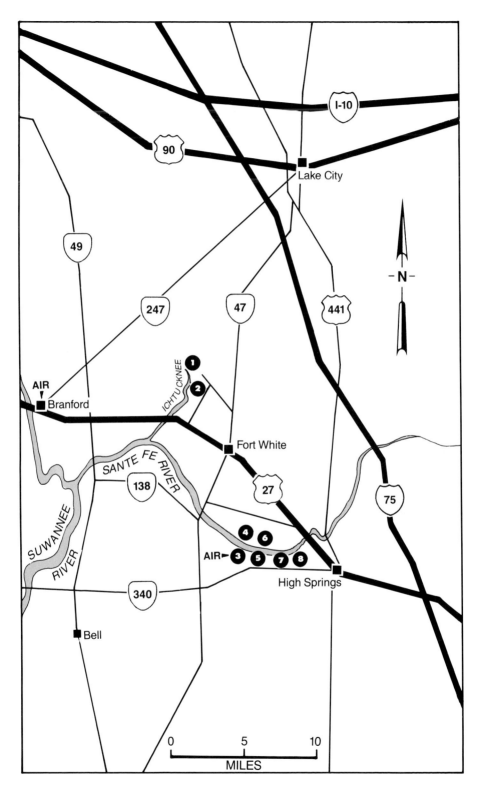

I-10

90

Lake City

49

247

47

441

ICHTUCKNEE

1
2

AIR
Branford

SANTE FE RIVER

Fort White

SUWANNEE RIVER

138

27

75

4 6

AIR 3 5 7 8

High Springs

340

Bell

0   5   10
MILES

Area Information

This area of north central Florida boasts the clearest diving waters to be found anywhere in the world. A beautiful series of springs lines the picturesque Santa Fe River, offering the diver a variety of large deepwater basins. Ginnie Springs, a commercially operated diving park, is not to be missed. Not only does Ginnie and its associated springs provide interesting diving, but the park also has restrooms with hot showers and great camping facilities.

One of the best ways to visit several of the springs in this chapter is to take a short canoe trip down the Santa Fe. The springs are easily spotted from the river. You can dive and then rest at each location for as long as you like before continuing downstream. The clear, captivating waters of the spring-fed Ichetucknee River rate high for underwater fun. The headsprings, as well as the three miles of the river, are operated and protected by the state.

Tanks can be filled at Ginnie. Gainesville also offers complete diving supplies and air. Canoe and raft rentals are easily found. Motels are located in Branford, High Springs, Lake City and Gainesville. The Ichetucknee Park has no camping areas, but facilities can be found at Ginnie Springs and Blue Springs on the Santa Fe River and in private campgrounds near the Ichetucknee park entrance.

NO.1 ICHETUCKNEE SPRINGS

Location: From the intersection of US 129 and 27 in Branford, go east about 7 miles. Turn left onto C-137 and travel 1.3 miles. Turn right and go 4.2 miles to the north park entrance.

OR

From the intersection of I-75 and S-47, travel south on S-47 for 12.5 miles. Turn right on S-238 by sign. Go 3.7 miles to the north park area.

Several springs, located in a beautiful woodlands, form the Ichetucknee River, a recreation paradise for both divers and nature lovers. The state nowoperates the springs and more than three miles of their run to the bridge on US 27. Park hours are from 8 AM until sunset. No camping is allowed inside the park area but excellent picnic facilities and restrooms are available.They have done a magnificent job preserving the area's natural beauty.

The three-hour float down the clearwater river remains unspoiled. There are no signs, houses, roads or concrete. The voyager will instead enjoy lush virgin woods filled with wildlife and the songs of birds. The headspring forms a circular pool 100 feet in diameter. The water emerges from a horizontal cavern near the north bank at a depth of 13 feet. The rate of flow is approximately 20 m.g.d., with a year-around temperature of 72 degrees. No scuba is allowed in the headspring.

The float down the run can be made on an assortment of items—anything from old tire tubes (the most popular item) to canoes. You drift slowly along in a two-knot current until you arrive at the US 27 bridge, two-and-a-half to three hours later. In the past, two cars were required for a group to make the trip. One car had to be left at the bridge to take you back to the headspring. Now, however, a bus service runs from the bridge to the headspring every hour (during summer only), thus eliminating the need for two cars.

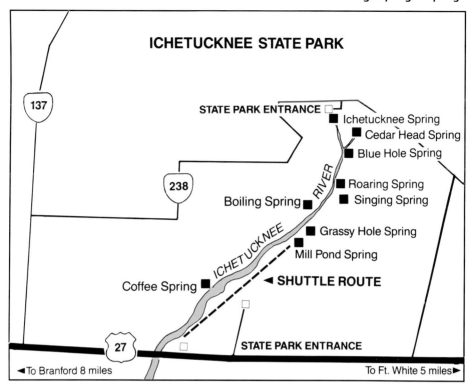

ICHETUCKNEE STATE PARK

137

STATE PARK ENTRANCE
Ichetucknee Spring
Cedar Head Spring
Blue Hole Spring

238

Roaring Spring
Boiling Spring
Singing Spring

RIVER

ICHETUCKNEE

Grassy Hole Spring

Mill Pond Spring

Coffee Spring

◄ SHUTTLE ROUTE

27

STATE PARK ENTRANCE

◄To Branford 8 miles

To Ft. White 5 miles►

Floating down the Ichetucknee.   Bill George

Divers enjoy snorkeling the run, studying the underwater beauty and searching the sands for fossils and relics. A wetsuit is recommended because of the long exposure to the cool water.

In 1983, the park opened an access to the river just below Mill Pond Spring. The entrance leading to a parking area is off US 27, a half mile east of the Ichetucknee Bridge. Now visitors have a choice of drifting the upper river section from the headspring to the northern section of the south park (1.4 miles), or from the northern section of the south park to the US 27 bridge (1.7 miles), or the entire stretch of 3.1 miles from the headspring to the US 27 bridge. The full run is open during the summer from June until Labor Day. The mid-point run is open year around. When you start down the run, remember there are no other roads or paths to leave the river by, and you will become exhausted trying to return against the current.

The new extended section of the south park provides a tram route to carry park visitors and a 1.8-mile pedestrian path. The state enforces a limit of 3,000 visitors daily—750 floaters are allowed on the upper section and 2,500 on the lower part of the river. It is advisable to arrive early, especially on summer weekends, if you plan to make the river trip. The park rangers have been assigned the duty of checking containers (including purses) for items that may cause litter in the river. No food, drinks or pets are allowed. Please cooperate.

NO.2 BLUE HOLE SPRING
(Restricted to Certified Cave Divers)
Location: Located inside the Ichetucknee State Park. A boardwalk beginning near the headsprings leads through the woods to Blue Hole. The 200 yard walk becomes quite long when you are carrying diving gear.

This is a beautiful spring that feeds the Ichetucknee River. The entrance, 15 feet across, is located near the center of the basin in about 13 feet of water. A shaft drops almost straight down 37 feet to a sand floor that widens to about 40 feet across. Diving into the shaft is made difficult by a strong flow. On the south side of the room there is a cave, seven feet in diameter. Its corridor is filled with a constant flow which makes penetration very difficult and tiring.

NO.3 GINNIE SPRINGS
Location: Going south from High Springs on US 41, turn west on S-340 (Poe Springs Road) and travel 6.6 miles to a graded road running to your right (Watch for the Ginnie Springs sign.). Go 1.2 miles to the entrance to Ginnie Springs grounds. Travel straight on small sand road until you arrive at the Ginnie Springs office. All guests must register here.

OR

From Branford travel east on US 27 to Ft. White. Turn right (south) on S-47 and go 9.9 miles. You will see a Ginnie Springs sign at this turn. Turn left on graded road and go 1.4 miles to the end; turn right and travel .7 mile to the entrance to Ginnie Springs. Turn left onto the small sand entrance road. Ginnie Springs office will be ahead on your right. Guests muust register here before going further.

Ginnie Springs, Devil's Eye Springs, Dogwood Spring and several smaller springs located along the banks of the beautiful Santa Fe River are now part of an outstanding camping and recreational facility. There is a charge to enter the area

GinnieSprings

R E S O R T

N ature is preserved here at Florida's number one freshwater dive site. This wilderness resort rests among 200 acres of unspoiled, northern Florida forest. Eight freshwater springs burst forth, forming beautiful basins that overflow into the pristine Santa Fe River.

Ginnie Springs is perfect for individuals, families or groups to relax in and have fun.

From camping (wilderness or with electricity), picnicking, swimming, snorkeling, canoeing, tubing and volleyball, to scuba lessons (beginner thru cavern and cave, and open-water training referrals) — or just browsing in the full-service dive center/country store for that one-of-a-kind momento. There are unlimited ways to have fun. Come experience Hidden Florida at its best.

For more information on rates, packages and discounts, call...

(800) 874-8571 • (904) 454-2202

Ginnie Springs, Inc. • 7300 NE Ginnie Springs Rd. • High Springs, FL 32643

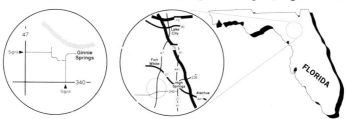

for diving or camping, but the improvements that have been made around the springs make it worth the price. The popular springs and their surroundings have been developed with much forethought.

Every convenience is available for the diver here, including strong cypress decks around the springs, two large bathhouses with hot showers, a 5,000 psi air station, scuba shop and complete line of rental equipment. Private campsites, with or without hookups, are available. A country store is located at the registration office. Without a doubt, Ginnie Springs is one of America's best diving resorts.

Ginnie's basin is a four-to-five-foot deep limestone shelf that suddenly drops 18 feet to a white sand floor. The entire area is bordered by waving eelgrass, creating a beautiful underwater setting. The cavern entrance, approximately four to six feet in size, runs horizontal to the sand floor. Just inside the mouth, the first chamber widens to 30 feet with a nine-foot ceiling. You next enter a large, sloping room approximately 60 feet wide and 70 feet long. The room angles downward from a depth of 35 feet to 60 feet, where it splits into two tunnels. The entrance into the narrow tunnels has been closed off with heavy grating.

NO.3 DEVIL'S EYE SPRING

Location: Found inside the Ginnie Springs grounds. Turn right just before the bathhouse and follow the sand road to the parking area. Cypress decks are located on both sides of the spring.

Devil's Eye is in the middle of a set of three springs located on the bank of the Santa Fe River. Its perfectly round shaft of limestone starts in six feet of water and drops straight down to a sand bottom at 20 feet. On the north side of the shaft, a cave entrance, three feet high and 18 feet wide, leads into a dark room called the "Devil's Dungeon". This room measures 30 feet long by 20 feet high. The limestone walls of the entire cave system are dark due to the mineral content of the water.

Certified cave divers can continue on to the north end of the room where they will find an extremely small cave measuring two feet by four feet. This dangerous corridor runs out under the river where a small room, about seven feet by ten feet, lies at a depth of 65 feet. Continuing a short distance farther, you'll see a shaft of sunlight. Follow the light up a steep rising chimney, about three feet by seven feet, until you enter the mouth of Devil's Ear, located at the edge of the river.

The third spring, called Little Devil, and located up the run from Devil's Eye, was recently dredged free of sand. The 40-foot long fissure is spectacular. Sheer walls drop over 45 feet. The view from the bottom is breathtaking.

NO.3 DOGWOOD SPRING

Location: Found within the Ginnie Springs grounds. The spring is 100 yards downriver (west) of Ginnie's Run.

Dogwood is a beautiful little spring with depths to 12 feet. There is good snorkeling in the spring and down the run. A deck has been constructed next to the spring for the convenience of visitors.

NO.4 JULY SPRING

Location: Directly across the Santa Fe River from Devil's Eye Spring. Swim across the river and up a short run bordered by lily pads.

July Spring is a small spring with a depth of 15 feet. The water flows from a long, narrow crevice. Because there are numerous springs feeding the Santa Fe in this area, a large section of the river becomes clear during periods of low water. As you glide down the river, its underwater habitat will unfold to reveal turtles, mullet, bass and bream hiding in a maze of sunken logs. The mud-filled crevices in the limestone floor conceal many relics and fossils. Slowly fan the silt from the cracks to reveal hidden objects.

NO.5 BLUE SPRINGS-SANTA FE RIVER
Location: Going south from High Springs on US 41, turn west on S-340 (Poe Springs Road) and travel 4.5 miles to Blue Springs sign. Turn north and follow the dirt road to the gate.

Blue Springs is commercially operated and an entrance fee is required. Although scuba is not allowed, it isn't really needed. Free-diving in the spring is great. A large limestone cliff drops 25 feet to a small cave entrance.

This is a good spring for underwater photography, combining excellent visibility with bold cliffs and aquatic plant life. The bream are tame and come out of the shadows by the hundreds when bread is offered.

A 1,500-foot boardwalk runs down the east side of the boil and follows its crystal run to the Santa Fe River. A small spring called Little Blue is located about 25 yards west of Blue.A sand beach borders Blue Spring on its south bank, and a swimming dock is located over the cliff drop-off. There are picnic tables and grills provided. Concessions are sold at the office. Good camping facilities are provided, and include a clean bathhouse.

NO.5 NAKED SPRING
Location: About 150 yards east of Blue Springs.

Naked Spring is 25 yards in diameter, with three small crevices lying about 12 feet below the surface. The water flows through thick woods and connects with Blue Springs' run before it reaches the Santa Fe.

In the olden days, the people who had swimming suits swam in Blue Springs, while those who couldn't afford trunks went back into the woods to swim at "Naked" Spring.

NO.6 RUM ISLAND SPRING
Location: From Branford travel east on US 27 to Fort White. Turn south on S-47 and go 4.2 miles. Turn left (east) on S-138. After 4.5 miles, turn right onto dirt road. Stay on primary dirt road for 1.4 miles until you reach the river. Continue on the road along the river for .2 mile to the spring.

A small spring located on the north bank of the Santa Fe River, Rum Island Spring fills a basin 40 feet in diameter with extremely clean water. Depths are to 12 feet. The spring is filled with a beautiful variety of aquatic plant life.

NO.7 LILY SPRING
Location: Go south from High Springs on US 41; turn west on S-340 (Poe Springs Road). Travel 3.4 miles and turn north onto dirt road. (Dirt road is located where the paved road changes texture.) Go .5 mile to a fork and bear left. Go .2 mile to the spring.

This small, picturesque spring area, located on the south bank of the Santa Fe River, is no longer open to the public. Plans for the area include the building of private homes around the spring basin.

NO.8 POE SPRING
Location: From US 27-41 in High Springs, go west on S-340 (Poe Spring Road) for 2.4 miles to a limestone road. Follow this road .7 mile to the spring.

This spring is fed by three small springs flowing down a 40-yard run to the Santa Fe River. It is ideal for family recreation, swimming and snorkeling. A crude boat ramp is available for launching fishing boats. The spring's temperature is 73 degrees and its flow is 45 m.g.d.

Blue Grotto   Ned DeLoach

BLUE GROTTO (Not on map)
Location: From the center of Williston (20 miles south of Gainesville) go 2 miles north on US 27A toward Bronson. Turn left onto the dirt road across from a Catholic Church. Follow the road 1/2 mile to the entrance on the right.

Blue Grotto was reopened in 1988 as a commercially operated diving park. The large limestone sink offers one of the state's best and safest cavern dives. Maximum depth is 110 feet. The water stays a constant 72 degrees and is incredibly clear except during brief periods in the hot summer months when the surface layer occasionally experiences an algae-bloom. When other spring sites in the area become dark because of rising rivers, Blue Grotto remains clear!

Diving couldn't be easier. There is an entry pool, floating dock and a guide rope around the cavern's interior. The site is on 15 acres that include a dive store with air and rentals, a bathhouse with hot showers and a picnic area.

SANTA FE RIVER CANOE TRAIL WITH SPRING SITES
(Big Awesome, Little Awesome, Trackone Siphon and Myrtle's Fissure)

Location: The best entry point is at Ginnie Springs.

The Santa Fe River is an unusual waterway. Born from a shallow central Florida lake, the small stream takes a northwesterly course through thick pine and hardwood timberlands before emptying into the Suwannee River. Along its winding route the river abandons convention, disappearing underground for miles at a time. The last resurgence occurs just north of High Springs as a placid

BLUE GROTTO

"One of the world's finest spring dives"

72 Degree temperature year round. Natural spring.
Two underwater platforms at 35 ft. in open water.
Open cavern with no tunnel maze.
Permanent heavy-duty guideline.
Fresh air bell at 30 feet.
Scuba and snorkel rental equipment.
Picnic tables beneath shady oak trees.
Dressing and rest room facilities.
2 Miles from Williston Center
with food and lodging.
Pure air fills to 5000 PSI.
Hot water showers.

**200 ft
Visibility**

**100 ft
Deep**

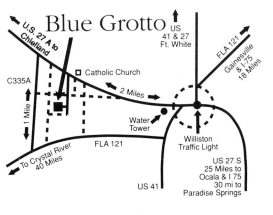

Route 3
Box 2790
Williston, Florida
32696
(904) 528-5770

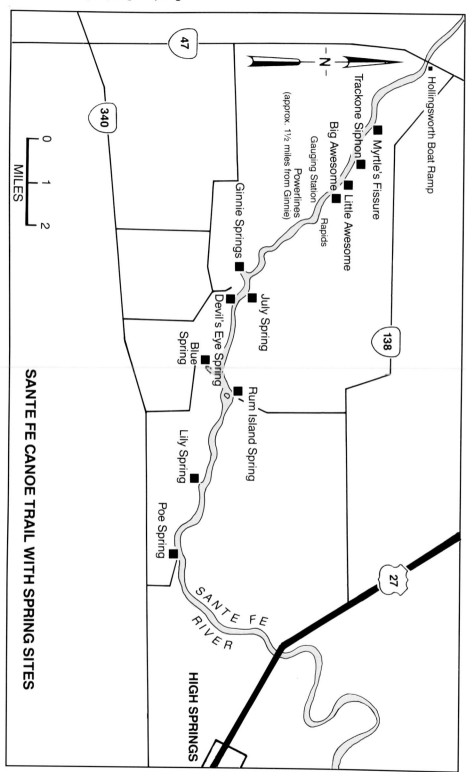

SANTE FE CANOE TRAIL WITH SPRING SITES

pool surrounded by cypress. From here her dark, tannic-stained waters are transformed by an influx of spectacularly clear water issuing from sister rivers that flow through limestone conduits deep underground. During periods of low rainfall these groundwaters dominate, causing the Santa Fe to run clear. This is when Santa Fe River diving is at its best.

The entire stretch from Ginnie Springs to the SR-47 bridge can be considered a dive site. Several interesting sinks and siphons are also located along the river's banks. When the river is low, visibility is 40 to 60 feet in most areas. Near the spring it is even better.

Fossil and Indian artifact hunting is great nearly everywhere. The sand-filled cracks between limestone slabs and siphons are good places to search. (Be sure to drag a tube supporting the dive flag to warn boaters of your underwater position.)

The river's abundant aquatic life can be observed with ease. Turtles and schools of mullet, bream, bass and gar are everywhere. Towering cypress and oaks line the banks. Another plus is the lack of home sites below the SR-47 bridge. The river runs calm except for a delightful set of rapids a mile below Ginnie.

Just after the rapids, on the north side of the river, is Big Awesome Slough, a salient river artery that forms an island. Here a large deep basin is found with depths of 40 feet. (Be careful to stay out of the cave here and at other spots mentioned. They are all dark, silty and dangerous.) A little further downstream you will spot a swirl of water near the north bank. Water is being sucked from the river into the aquifer. This is Little Awesome Siphon. A few hundred yards further downriver in front of a low, white bluff is Trackone Siphon. There is a large pool here with depths to 60 feet.

There are two more small spring-siphons before you reach Myrtle's Fissure. Up a shallow run, 20 yards from the river, limestone walls plunge to 60 feet. Gently deflate your b.c., carefully avoiding the silty walls, drop to the floor, and roll over and look up. You will never forget the sight.

A short distance downriver is the SR-47 bridge. A boat ramp is located just past the bridge on the north bank. From here to the Suwannee the diving is less interesting and there are many homes on both sides of the river.

Those who would like more time on the Santa Fe should put in on the north side of the US 27 bridge. This is located 3 miles NW from the town of High Springs and approximately six miles by river from Ginnie Springs. This, too, is a beautiful stretch of the river with several spring sites. Poe Spring (No.8) is near the south bank, 1.7 miles below the bridge. One mile downstream is Lily Spring (No.7). Rum Island Spring (No.6) is a little over a mile from Lily. It is at the north end of the only island on this section of the river.

The clear run from commercially operated Blue Springs (No.5) is less than a half mile downriver. Canoeists are not allowed to enter the spring basin. The Ginnie Springs Corporation property begins two miles further downstream. You will easily spot Devil's Ear and Eye Springs on the left bank. July Spring is just across the river. The run to beautiful Ginnie Springs is less than a quarter mile from here.

3

CENTRAL FLORIDA

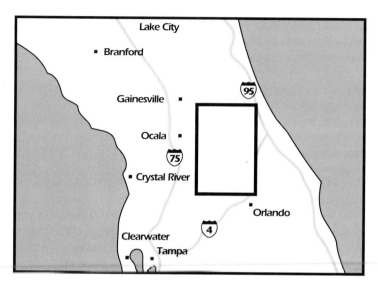

Spring Diving Locations

1. Devil's Sink
2. Orange Spring
3. Croaker Hole
4. Salt Springs
   Clearwater Canoe &
   Snorkel Trail
5. Silver Glen Springs
6. Juniper Springs
   Clearwater Canoe &
   Snorkel Trail

7. Alexander Springs
   Clearwater Canoe &
   Snorkel Trail
8. DeLeon Springs
9. Blue Spring-Orange City
10. Rock Springs
11. Wekiwa Springs
12. Apopka Spring
NOT ON MAP:
   Kingsley Lake Spring
   Forty Fathom Grotto

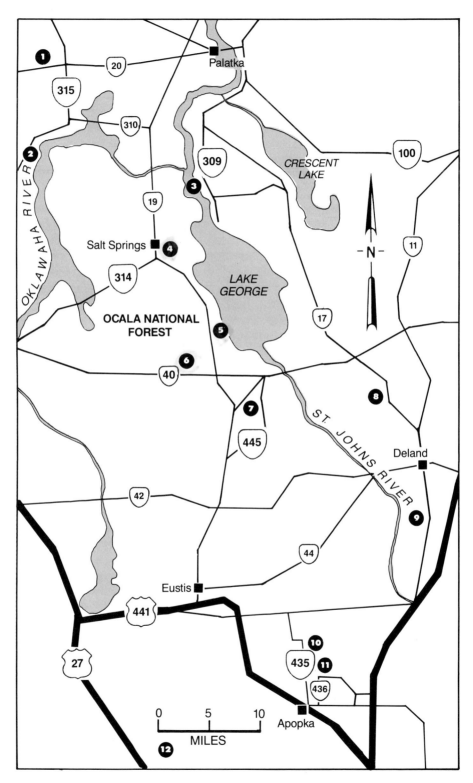

Area Information

Several of the state's most beautiful and accessible inland diving areas are found in and around the Ocala National Forest in central Florida. The most popular diving locations are commercially operated and have varying requirements for scuba diving. Juniper Springs, Alexander Springs, Blue Spring, DeLeon Springs and Salt Springs are operated by state or federal agencies. Scuba is allowed in Alexander Springs and Blue Spring for certified divers. Canoeing is popular, and canoe rentals are available at Blue Spring, Silver Glen Springs, DeLeon Springs, Juniper Springs and Alexander Springs. Air and divers' supplies are available in Gainesville, Orlando, Daytona Beach, DeLeon Springs and DeLand.

For hikers the 60-mile Ocala National Recreation Trail stretches from the Clearwater Lake Recreation Area, located south of Alexander Springs on C-38, to the Rodman Dam at the northern edge of the Forest. Juniper and Alexander Springs have short, interpretive nature trails.

NO.1 DEVIL'S SINK
Location: From Gainesville, take S-20 east through Hawthorne. Where S-21 turns south to Johnson go straight for 1.9 miles on S-20. Turn left on the paved road that runs up the side of the hill (road is across from the lake). Go about 100 yards and take one of the two small sand roads running through the woods on your left. Follow the road back to the sink.

Devil's Sink is a spectacular sight with 75-foot bluffs that plunge almost straight down to water level. There is a path on the south side which leads down. The sink widens into a large room a short distance beneath the surface and drops to a depth of 80 feet.

NO.2 ORANGE SPRING
Location: In the town of Orange Springs. From S-21 south of Orange Creek bridge, go east on dirt road. Then turn north on next intersecting dirt road and go .2 mile to spring.

Orange Spring contains some of the largest sand boils available to divers. The boil issues 6 m.g.d. from a sandy bowl near the middle of the pool. It is always clear and reaches a depth of 25 feet. The area is not visually impressive and the water contains a considerable amount of sulfur.

NO.3 CROAKER HOLE
Location: From Palatka, go south on US 17 for 10 miles to S-309. Turn right on S-309 and continue for 10 miles to Gateway Fish Camp (4 miles south of Welaka). You can rent or launch a boat from there and go 1.5 miles downstream (north) to Little Lake George. Croaker Hole is near the southwest corner of the lake, about .2 mile north of Norwalk Point. The spring is easily located in calm weather because of a "slick" formed on the surface. Fishermen have sunk poles into the bottom around the spring to tie their boats up.

Tie off a safety line and dive with a light directly into the boil's center. At a depth of 20 feet you'll pass through the terrible lake visibility and into clear spring water. At 40 feet there is a horizontal cave entrance, 25 feet by 12 feet. By pulling your way in against the strong current, you can swim into an exceptionally picturesque tunnel for about 100 feet before it ends in a tumble of boulders. The flow enters from several narrow crevices around the boulders. One of the crevices issues salt water, creating eerie lighting effects called a halocline.

There is absolutely no surface light coming through the dirty water over the spring. Lights and line must be used. This is a popular fishing spot so be cautious of fishing lines. Always carry a knife in case of entanglement.

NO.4 SALT SPRINGS
Location: Just north of the intersection of S-19 and S-314 in the Ocala National Forest at the small community of Salt Springs.

The Salt Springs area is managed by the US Forest Service, but operated by a private concessionaire. The grounds are open at 8 a.m. and closed at 8 p.m. A entrance fee of $1.50 is charged. There is no camping here but the Forest Service operates a camping area a half mile north on S-19. No scuba is allowed but snorkeling is excellent in the large, 100-foot diameter, spring pool. A concrete retaining wall and boardwalk borders the clear water on three sides. A towering stand of oaks draped with spanish moss dominates the well-maintained grounds.

For the most part, the basin is shallow, four to six feet, but there are several limestone fissures that drop to 15 feet. Grass beds surrounding the limestone vents harbor a variety of interesting aquatic creatures. These waters are well-known for the large number of blue crabs. Water temperature remains a constant 75 degrees year-around. The three main vents issue 53 m.g.d. of sparkling water that feeds a

wide, placid four-mile run to Lake George. Swimming and snorkeling are restricted to the basin. Canoes can be rented at a nearby marina.

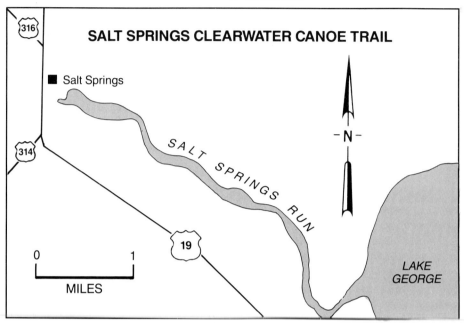

SALT SPRINGS CLEARWATER CANOE AND SNORKEL TRAIL

The wide, clear run starts at the spring basin at the town of Salt Springs. It flows 4 miles southwest to Lake George. There is not a good take-out point at the lake so you must canoe back to the spring. The run has virtually no flow so plan on paddling the entire distance. The stream is filled with aquatic life. Mullet, bream and blue crabs are everywhere. Bird watching is excellent in this remote area.

NO.5 SILVER GLEN SPRINGS

Location: From the intersection of S-19 and S-40 in the Ocala National Forest, travel 6.4 miles north on S-19. Turn east on a dirt road that runs under a large wooden archway (sign by the road).

This is a large spring area with two small caves issuing a flow of 72 m.g.d. The cave on the spring basin's north side has recently been opened to swimmers and snorkelers. A circular cave entrance 7 feet across and 8 feet below the surface opens into a large room with depths to 42 feet. It is a challenging free-dive to plunge straight through the outflow into the cave. Schooling striped bass often swim just inside the entrance. The main spring cave, located at the end of a wooden dock, is at the base of a ledge in 28 feet of water. No scuba is allowed at either spot.

Silver Glen is a wonderland for the snorkeler. Hours can be spent exploring its grass beds and white sand floor. Fish are plentiful and easily approached. Relics have been found in the half-mile run to Lake George. It is a good area for family recreation, with a white sandy beach and a section roped off for swimmers. A boat and canoe rental service is provided at the office. The area also boasts four-and-a-half miles of wilderness camping along the lake, with several Indian shell mounds dating back 400 years. Sixteen small sand boils, called the Laughing Sands, are located along a small path which runs north from the spring. Cabins are available for rent, and there are many trailer and tent camping sites around the grounds.

Silver Glen is a privately owned recreational area with the largest marina on Lake George. For further information or reservations call (904) 685-2514.

NO.6 JUNIPER SPRINGS

Location: From the intersection of S-19 and S-40 in the Ocala National Forest, travel west on S-40 for 4.7 miles to a sign at the entrance.

Juniper Springs is one of the oldest National Forest Recreational Areas in the east. It was built in the mid-1930s by the Civilian Conservation Corps. Today, the popular area has 79 camping units with three clean restroom facilities. There is also a self-service laundry and a canoe concession.

Juniper Springs is a large concrete-bordered pool 135 feet long and 80 feet wide. The water stays at a constant 72 degrees. No scuba is allowed but the snorkeling is good. The pool's shallow, grassy bottom gently slopes to the eastern end where it drops to 16 feet at the spring vent. The water's flow is channeled by a scenic millhouse before widening into a spring run.

Fern Hammock Spring is located less than a quarter mile west. Its flow joins Juniper's to form Juniper Creek, one of Florida's favorite clear water canoe trails. There is no swimming or fishing allowed at Fern Hammock.

JUNIPER CREEK CLEARWATER CANOE AND SNORKEL TRAIL

This 10-mile canoe trail starts at the Juniper Springs Recreational Area and flows

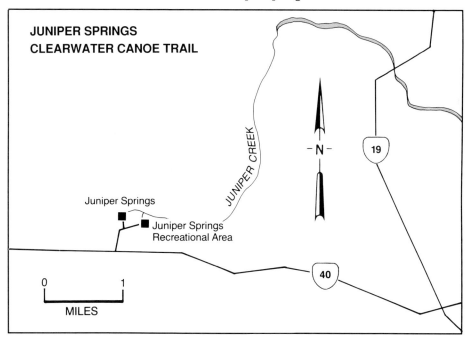

JUNIPER SPRINGS
CLEARWATER CANOE TRAIL

JUNIPER CREEK

- N -

19

Juniper Springs

Juniper Springs
Recreational Area

40

0   1

MILES

through semitropical lowlands to Lake George. A canoe rental service is available at the Recreational Area. They also provide a shuttle to bring you and the canoe back to Juniper. The pick-up spot is 7 miles downstream, where the creek passes under S-19. There are no other accesses to the creek before this point. This is a popular canoe trail; reservations are advised. For reservations and information call (904) 625-2808 between 9 a.m. and 12 p.m.

The creek is narrow for the first two miles. Both banks are an impenetrable tangle of tropical growth that supports many species of wildlife. Bird watching is always good. The creek is filled with an abundance of fish and turtles. It is rare if alligators are not sighted during your trip.

A picnic area called Half-Way Landing is 3.6 miles downstream. After this point the waterway widens as it flows through a grassland area.

The pick-up spot is on the southeast side of the bridge over S-19. This is reached by driving east on S-40 for 4.7 miles. Turn left (north) on S-19 and go 3.8 miles to the bridge.

The creek continues southeast for 3 miles to Lake George, but there is not a pick-up point where the stream enters the lake.

NO.7 ALEXANDER SPRINGS

Location: From the intersection of S-19 and S-40 in the Ocala National Forest, travel east on S-40 for about 5 miles. Turn right at Alexander Springs sign, and go a short distance; turn right at the stop sign. After .5 mile, turn left on S-445 and go about 6 miles to the sign at the entrance.

Alexander Springs is a National Forest Recreation Area. There is an entrance fee, and all persons using scuba are checked for certification. The huge spring basin is fed 96 m.g.d. of clear water from a single cave located at a depth of 27 feet. The area surrounding the boil provides good snorkeling. A large covering of lily pads hides hundreds of bream, mullet and bass. A beautiful 15-mile run flows to the St. John's River and provides good relic hunting. It is an excellent camping area with complete trailer connections and numerous tent sites.

For camping information call the Seminole Ranger District (904) 357-3721.

ALEXANDER SPRINGS CLEARWATER CANOE AND SNORKEL TRAIL

Although the clear creek winds west for 15 miles to the St. Johns River, canoeists generally paddle only the upper 6.5 mile section. There is a canoe rental in the Recreational Area but there is not a shuttle service available. You start at the Alexander Springs basin. The spring creek is wide and slow moving. Both banks are covered with thick tangles of subtropical growth. After a distance of 1.4 miles you pass under the C-445 bridge. In 2.7 miles you pass a few buildings on the south shore. This is Ellis Landing. At this point the river narrows and winds around several islands.

The take-out is at a boat ramp on the north bank. To get there by car travel east on C-445 for one mile. Turn right (east) on C-52. This is the first road past the bridge. Go 1.7 miles before turning right (south) onto C-52B. This road goes 2.7 miles before ending at the boat ramp.

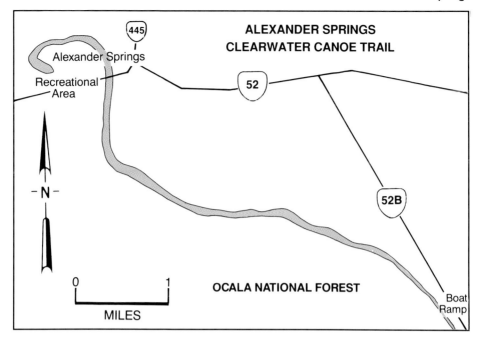

ALEXANDER SPRINGS CLEARWATER CANOE TRAIL

OCALA NATIONAL FOREST

NO.8 DELEON SPRINGS

Location: Just west of the town DeLeon Springs, which is 9 miles north of DeLand on US-17.

DeLeon Springs is located in a State Recreational Area. There is a 50 cent admission fee. The large spring basin is 170 feet across. It is bordered by a concrete restraining wall and walkway. The well-maintained grounds are shaded by a large stand of oaks dripping with Spanish moss. There is a restaurant; snack bar and canoe rentals are available.

Scuba is only allowed with instructors who pay an annual fee to use the basin for classes. A small cave entrance is located in 30 feet of water near the basin's center. Nearly 20 m.g.d. flow from this single opening. The cave's maximum depth is 40 feet and can be penetrated for 170 feet. The open spring pool has a shallow sand floor patched by large beds of aquatic vegetation. Spring Garden Creek flows from the headpool 10 miles east through a series of three lakes to the St. Johns River. This entire area is known for the wealth of Indian artifacts and fossils. L.L. "Lucky" McKee, owner of the DeLeon Springs Dive Shop, has discovered many of the relics he displays at his business.

NO.9 BLUE SPRING—ORANGE CITY

Location: From DeLand travel south on US 17/92 for 6 miles to Orange City. Turn west at a traffic light onto West French Ave. Watch for a large overhead sign at this intersection. Go west for 3.2 miles. Turn left on the first road past a railroad overpass. The park check-in station is .2 mile down this road.

Blue Spring is known for its cave dive and manatees. The large aquatic mammals come up the spring run from the St. Johns River during the cool winter months for the warmer 72-degree spring water. These endangered animals are closely protected by the park rangers. You are not allowed within 50 feet of the manatees while underwater or in a canoe. You will have to be satisfied to watch them from a distance, or from the boardwalk that follows the run's eastern shoreline.

In order to dive Blue Spring, a visitor must have an up-to-date certification card. You must check in at the park office and register to dive before 3 p.m. Each diver must have a full tank, an underwater light and a diver's knife. If you plan to dive on the weekend, arrive early. The park allows only 40 divers in the basin at one time.

To get to the basin follow the boardwalk north past the concession area. There are steps leading into the water. This is not a dive for beginners or those who are in poor physical condition. The cave is deep and demanding. A strong outflow must be negotiated during the entire dive.

The cave is located in 10 feet of water at the run's north end. It drops straight down to 60 feet. There are several dead-end tunnels that lead off this shaft. All are small and silty and should be avoided. At 60 feet the cave's shaft angles at 45 degrees and continues to a restriction at a depth of 120 feet. The flow here is extremely strong.

Camping is excellent at Blue Spring (except for the close proximity to the railroad tracks). There are 51 campsites; 27 with hook-ups. Six vacation cottages are available. Each sleeps eight.

There are several marked canoe trails from 6 to 14 miles in length. They follow the .5 mile run to the St. Johns River where they wind around islands and down old logging canals.

For reservations or current information about diving call the Blue Spring State Park office at (904) 775-3663 between 8AM and 5AM, Monday through Friday.

NO.10 ROCK SPRINGS
Location: From Apopka, go north on S-435 for 5.8 miles to the dead end at Bay Ridge-Rock Springs Road. Turn right (east) and travel .4 mile to Kelly Park entrance and the springs.

This spectacular spring emerges from the dry cave at the base of a high limestone cliff. Unfortunately, a grating prevents access into the spring source. The 1.5-mile run is picturesque, and has many plants and fish which provide hours of interesting snorkeling.

NO.11 WEKIWA SPRINGS
Location: From the town of Apopka, travel east on US 441 for 1 mile. Turn left on S-436 and go 1.5 miles. Then turn left on the paved road, and go 2.9 miles to sign on left of road.

The spring is now part of a state park. No scuba is allowed. There are two springs in a kidney-shaped pool which measures 120 feet in diameter and attains a depth of 20 feet. The flow is measured at 44 m.g.d. and the water temperature is 75 degrees. The run is one of the headwaters of the Wekiwa River, a tributary of the St. Johns.

NO.12 APOPKA SPRING
Location: In the southwest section of Lake Apopka, near a narrow arm of land called "Gourd Neck." From the small town of Oakland (about 10 miles west of Orlando), go west for 3.7 miles on S-438. Turn right onto a sand road that runs through an orange grove. Continue for .5 mile until the road ends. The spring is 150 yards north

of the road's end. It might be best to go by boat from the Gourd Neck Fish Camp. Murky surface water nearly always covers the spring area, so watch for the boil.

The soft, debris-strewn lake bottom slopes gradually to the spring vent at a depth of 35 feet. Since the water may not be clear until you reach the cave, it is best to tie a line to a float in the center of the boil and descend straight down into the spring. The narrow (two feet by three feet) spring vent opens into a beautiful sloping cavern, 25 feet wide and 50 feet long, and goes to a depth of 80 feet. This must be considered an advanced cave dive because of the complete absence of light, the debris covering the surface, and the large amount of lost monofilament line and hooks.

KINGSLEY LAKE SPRING (Not on Map)
Location: In Kingsley Lake in Clay County, just out from Strickland's Landing. Travel east on S-16 from US-301 to the small town of Kingsley Lake. Watch for signs.

This spring issues clear water from a depth of about 45 feet. The spring opening is too narrow to enter. The lake itself is reasonably clear, with visibility hovering near 20 feet and has a depth falling off evenly to 85 feet in the middle.

FORTY FATHOM GROTTO (Not on Map)
Location: Travel north from Ocala on I-75 to Exit 71. Go west on S-326 for 7 miles to NW 115th Street (a dirt road) and turn right. This road is just past the Cut Above horse ranch. Follow the dirt road for a quarter mile. The grotto is just off to the right. Anyone interested in diving the grotto should first contact Hal Watts' Scuba Plus, 2219 E. Colonial Dr., Orlando, FL (407) 896-4541.

Forty Fathom Grotto was discovered as a diving site by Hal Watts and Bob Brown in 1962. When first dived it was known as Zuber Sink, but the name was changed to Forty Fathoms because of its 240-foot depth. However, access to sport diving has been limited for years because of ownership problems. In 1986 the eight-acre tract surrounding the grotto was purchased by Hal, who is offering access to the public through his diving business in Orlando.

Wooden steps lead down a 70-foot slope to water level. A large floating dock makes water entries a snap. The oblong pool is 150 feet wide and 225 feet long on the surface, but opens underwater into a larger area 200 feet by 400 feet. Visibility most of the year varies from 50 to 75 feet. Bottom depths are from 90 to 240 feet. The walls of a large ledge at 40 feet are covered with ancient sand dollars and sea biscuits dating back 30 million years to the Ocala Limestone Era. There are two small caves at 50 feet and an extremely dangerous opening at 100 feet. Absolutely no diving is allowed here.

There is interesting diving on the wreckage of twelve cars and two motorcycles resting on the slope from 90 to 225 feet. The oldest is a 1928 Chrysler. A 1968 Corvette can also be explored. A diver's platform at 30 feet is provided for the convenience of entry level check-out dives.

The entire site is fenced and kept locked. Diving is allowed only with guides from the Orlando diving school. Other dive stores may bring groups by reservation after their leader has been checked out in the grotto.

This is a good location to teach U/W navigation, light salvage and basic deep diving, which is limited to 130 feet. Parking is convenient and there is plenty of space for cookouts and topside recreation.

4

WEST COAST

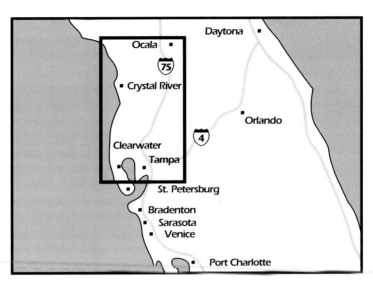

Spring Diving Locations

1. Rainbow River
2. Paradise Springs
3. Nichols Spring
4. Crystal River
5. Chassahowitzka
   River Springs
6. Joe's Sink
7. Weeki Wachi Run
   Hospital Hole
8. Saw Grass Spring
9. Palm Sink
10. Crystal Springs
11. Lithia Spring

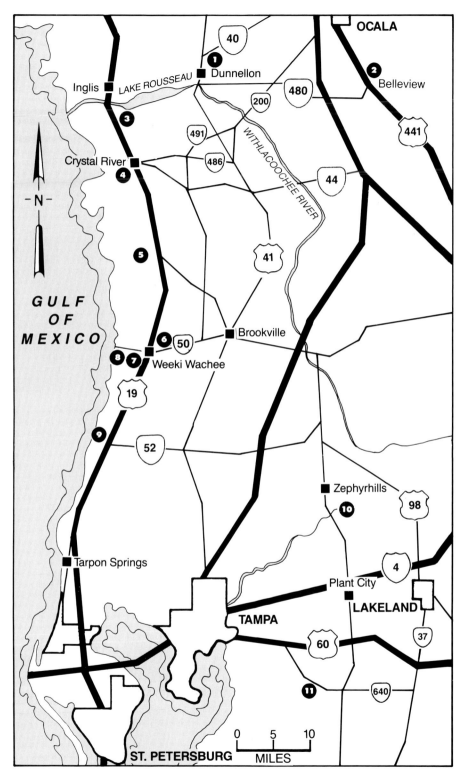

GULF
OF
MEXICO

OCALA

40

❶ Dunnellon

❷ Belleview

Inglis LAKE ROUSSEAU

480

200

441

❸

491

WITHLACOOCHEE RIVER

486

44

Crystal River

❹

41

❺

Brookville

❻ 50

❽❼ Weeki Wachee

19

52

Zephyrhills

❿

98

❾

Tarpon Springs

Plant City

4

LAKELAND

TAMPA

60

37

⓫

640

0   5   10

ST. PETERSBURG MILES

-N-

103

Area Information

Large springs, sinks and spring-fed rivers are common along Florida's west coast. Good diving conditions generally remain constant throughout the year. However, a few sinkholes occasionally experience algae growth during the summer months. Crystal River, located just off US 19, is one of the world's most popular freshwater dives. Water visibility is good throughout the year, providing exciting diving for both novice and experienced divers. Air and divers' supplies are available in Crystal River, Tampa, St. Petersburg, Clearwater, Ocala, Zephyrhills and Lakeland. Spring-fed Rainbow River and Chassahowitzka River must also be considered as two of Florida's best freshwater diving attractions.

NO.1 RAINBOW RIVER
Location: From the US 41 bridge over the Withlacooche River travel N through the town of Dunnellon for 2.5 miles. Turn right (east) on the paved road SW 90 LN (watch for small brown P.K. Hole Park sign). Go 1.2 mile before turning left just over RR tracks. Continue .7 mile to park entrance gate on the right. The road leading to the boat ramp is the next immediate right.

The small county park charges a $1.00 entrance fee. There is a small concession stand that has snacks and rents tubes for the float to the Withlacoochee River; however, no canoe or boat rentals are available for the trip upstream to the spring basin. A launching fee of $1.00 for boats and $.50 for canoes is paid at the concession. The park is open 8 AM till sunset in the summer and 9 AM till 5 PM in the winter.

This is a large, clear river that runs for five-and-a-half miles from Rainbow Springs to the Withlacoochee River. The area around the headsprings (once a tourist attraction) is closed to the public. The headsprings can be reached by boat, which can be launched from the boat ramp at the recreational area. It is a mile and a quarter trip upriver to the springs.

Rainbow Springs is a large basin fed through many sand boils. Depths range from 7 to 30 feet. The water is crystal clear, making this a good area for underwater photography. After diving the headsprings, you can drift back downriver with your boat, and either snorkel or use scuba. There are several small springs located in the river bottom. Relic hunting is good and fish are abundant. Depths range from 10 to 20 feet.

NO.2 PARADISE SPRINGS
Location: From the intersection of S-40 and US 27/441 in Ocala travel S on US 27/441 for 8 miles. This is a divided highway. Where the center divide of trees ends (look for a tall On Top of the World billboard in the median facing south),cross over to the north bound lane and go .3 mile to a small sand road (watch closely to the right for a large mailbox with a painted divers' flag). Follow the sand road for .5 mile making a sharp right just over the RR tracks. Contact the owners in the yellow house.

Paradise Springs is actually a sinkhole formerly known as Archway Sink and before that Wolf Sink. The new owners have gone to considerable effort to transform the site from a unsightly dump area into a pleasant and quite interesting

Paradise Springs

When you enter the crystal clear, 74° water, you are immediately surrounded by evidence of Florida as it must have been when it was part of the sea, millions of years ago.

Over 10,000 years old, thousands of sand dollars, sea biscuits, shells and bone fragments remain imbedded in the ceilings and walls of this unique diving experience.

- Large open cavern
- 200 Ft. Visibilty
- Over 100 Ft. Deep

For discriminating divers wishing to experience this natural wonder first hand, reservations must be made at least 24 hours in advance by calling

(904) 368-5746

Jim and Nancy Paradiso, Managers

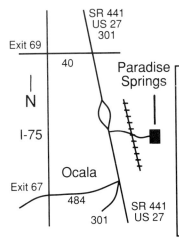

FROM THE NORTH

Take I-75 south to exit 69 (SR 40). Go east 3 miles to SR 441 (Pine Ave) and turn right. Follow about 8 miles to where the road splits (trees in the center). Take the first U-turn after the road comes together. Now heading north on 441, where the road starts to divide you will see a black mailbox and a dirt road on the right. Follow dirt road 1/2 mile to Paradise Springs.

FROM THE SOUTH

Take I-75 north to exit 67 (SR 484) and proceed east app. 7 miles. Turn left on SR 441 and go five miles to where the road splits (trees in the center). 50 feet before the split you will see a black mailbox and dirt road on the right hand side. Follow dirt road 1/2 mile to Paradise Springs.

cavern dive. The 40-foot slope to the small surface pool of clear water is lush with ferns and native trees. A walkway of natural stone now leads to a wooden deck at water's edge.

The small 20-foot diameter pool immediately opens into two large cavern rooms. The northern room is 40 feet wide but closes off 60 feet back where the water vents into the chamber. The main cavern to the south is 60 feet wide with a 45-foot clearance. Eighty feet back the passage narrows slightly and slopes down at a 45 degree angle to a depth of 100 feet. From here a narrow tunnel drops to a depth of 140 feet.

The passages run through an extensive fossil bed. Sand dollars and sea biscuits cover the walls of the first room. Large bones of yet unidentified species and a set of petrified ribs extend from the wall. Reservations are required. They can be made by calling (904) 368-5746. Divers should be cavern or cave certified.

NO.3 NICHOLS SPRING

Location: From US 19 south of Inglis, turn east on the first paved road south of the large group of power lines which run just south of the barge canal. The road angles sharply back north. Follow signs to boat ramp at the power dam.

There are actually two springs located in the river between the boat ramp and the dam. Their boils can be spotted easily from the shore. The closest spring is Nichols; the other is called Fire Hydrant Spring. Both provide interesting diving. Numerous fossils have been recovered from around the spring area and down the river.

NO.4 CRYSTAL RIVER

Location: In the town of Crystal River along US 19, which runs down Florida's west coast.

Crystal River is one of the most popular freshwater dives in the world. Although there are 30 known springs and sinks in the King's Bay area, only eight sites are of consequence to the diver. The springs are all located in King's Bay and connecting canals. A boat is needed to dive in the bay and canals. Boats can be rented from one of several fully-equipped diving concessions on the bay. It is only a short boat trip to the springs from any of the shops. As the springs are little affected by the wind, rain or tides, the diving is always good.

The largest and most popular spring is King's Spring, just off the south bank of Banana Island. Anchor your boat in four feet of water that surrounds the spring. The spring is 75 feet across, dropping almost straight down to 30 feet, where you will find two entrances to a cave that goes to a depth of 60 feet and back about 50 feet. Inside the cave, your eyes quickly adjust to the dimness, and you'll have no problem finding your way around. The sunlight cuts its way through the clear water and streams inside the cave, creating a breathtaking sight. The underwater photographer will be in paradise among the bold cliffs and crevices. Silhouette shots are spectacular. Fish are always on hand and easy to approach. Many varieties of fresh and saltwater fish are common to the river. In winter they are attracted by the warm 72-degree water. In addition to bass and bream, the spring is filled with mullet, trout, redfish, sheepshead, gar, snook, tarpon and snapper. Spearfishing is illegal.

Just west of King's Spring is Grand Canyon Spring. It is a 35-foot-long crack in the rock at a depth of 25 feet. A strong flow issues from a three-foot hole at the west end of the canyon.

Mullet's Gullet, located 100 feet east of King's Spring, is a series of several small springs in 20 feet of water. The flow comes from crevices in the smooth rock bottom. The clear water of the area and shallow depths make this an ideal locale for the underwater photographer to get some great shots.

Shark Sink is located on the west side of the bay, about 100 feet out from a dock running off a land point. Two holes angle down to 45 feet. A strong flow comes from the bottom of the springs. There are no caves. The area around the holes is shallow and thick with aquatic plants.

Idiot's Delight is a group of three vertical shafts located in a canal system on the east side of the bay. The largest shaft is about four feet in diameter. The water is unusually clear around the springs, and fish are plentiful. The land surrounding the springs is private, so stay with your boat.

Three Sisters Spring is a group of five springs situated a few hundred feet north of Idiot's Delight. Depths range from 18 to 25 feet.

Gator Hole is located in the canal system on the east side of the bay. Go under the bridge and to the second canal on the left, just out from a concrete dock. Gator Hole is the entrance to an extensive cave system that reportedly collapsed in 1963, shutting off the passage. You can still dive into the entrance room to a depth of 35 feet. This is a good place to spot manatees in the winter, and there is always an abundance of fish around.

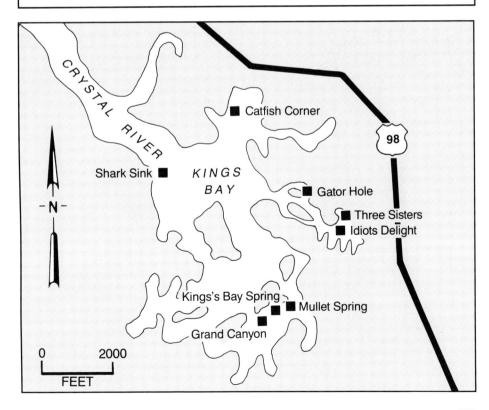

Catfish Corner is near the shore on the north side of the bay, just out from a wooden dock. A hole, four feet in diameter, starts in 15 feet of water and opens into a 10 by 20 foot room. It is home to hundreds of catfish of various sizes. An underwater light is not necessary, but the dive will be more enjoyable if you bring one along.

Crystal River is famous for its herds of sea cows (manatees) that come upriver in December and stay around the springs throughout the winter. One winter there were more than 40 of the huge, warm-blooded mammals counted. The average cow measures eight to ten feet in length and is perfectly harmless. A sea cow is very timid, especially when several divers are in an area. Because they are scared of scuba, approach them only while snorkeling. Swim slowly over the animal and wait for him to come up to you. Be gentle: if the manatee doesn't come to you, be satisfied to view him from a distance.

The manatee was officially designated an endangered species in 1967. Laws against harassment of the animals are strictly enforced by the state's Department of Natural Resources in the King's Bay area between November 15 and March 31 of each year. Harassment is defined in the Endangered Species Act as "An intentional or negligent act or omission which creates the likelihood of injury to wildlife by annoying it to such an extent as to significantly disrupt normal behavioral patterns which include, but are not limited to, feeding or sheltering." Aggressive pursuit of the manatee by a boat or diver affects the normal behavior of the animals and may cause injury to them by exposing them to cold water temperatures if they are forced to flee. This would be considered a violation of the Endangered Species Act. It is most important to go slow in your motorboat and keep a sharp lookout for manatees. They like to rest just under the water's surface. A mistake on your part could cause serious mutilation or even death to the mammals.

NO.5 CHASSAHOWITZKA RIVER SPRINGS

Location: 14 miles south of Crystal River, or 7 miles south of Homosassa Springs. From US 19, turn west on S-480. Follow the road for 1.9 miles as it curves through the small town of Chassahowitzka and ends at a fish camp.

The clear Chassahowitzka (Chassa-witz-ka) River, fed by more than a dozen freshwater springs, is one of Florida's most beautiful and unspoiled wildlife refuges. The shallow-water stream meanders through a primitive cypress and hardwood forest on its way to the Gulf. Numerous spring-fed creeks branch off into forgotten woodlands. Divers, canoeists, nature lovers, and adventurers will be spellbound by its majesty. Bass, bream, mullet, catfish, turtles, eels, crabs, egrets, herons, pelicans, ospreys, hawks, cormorants and wild ducks live here in profusion.

The main spring, Devil's Punch Bowl, is located about 50 yards from the fish camp. It has depths to 30 feet and a constant 74-degree temperature.

From the main spring, continue downstream for 200 yards until you spot the large Crab Creek run on the right. Continue up the run 300 yards to Crab Creek Springs, a series of three saline boils in scenic limestone shafts. The flow emerges from openings too narrow for entry. Depths are to 20 feet with visibility about 40 feet most of the year. There's great crabbing for large blue crabs in the eel grass along the run.

Two hundred yards downriver from Crab Creek is Houseboat Springs. A series of crystal clear boiling springs is found in a 200-foot diameter bay adjoining the river on the left (south) bank. There is picturesque diving with large schools of fish in the underwater hydrilla forest surrounding the springs. Maximum depth is 20 feet.

Blue Springs is at the head of half-mile-long Blue Creek. The mouth of Blue Creek enters the river 200 yards downstream from Houseboat Springs. The narrow, fast-moving run winds its way through a subtropical jungle teeming with sights and sounds of wildlife. The large Blue Springs basin is 25 feet in depth. Clear water flows from numerous tiny cracks in the bedrock floor. Visibility is limited to 30 feet. The cloudiness of the blue water is caused by a form of brown algae. Nevertheless, the spring is quite an interesting dive. There is a large sunken boat to explore. The good people at the fish camp take canoeists to Blue Springs for the start of an easy, scenic, four-hour trip back to the main spring area. What a way to spend a lazy summer afternoon in Florida!

Uncle Paul's Sink was discovered by divers in 1982. It is located in a small bay on the west side of the run, 300 yards below Blue Springs. It is a dramatic fissure, 25 feet long by 10 feet wide, that makes a sheer drop from five to 60 feet. The side walls spread slightly as the diver descends. Visibility is about 40 feet in the saltwater below the upper ten feet of cloudy fresh water.

Three Sisters Springs is made up of three beautiful, freshwater boils in a clear run 100 yards upstream from the main spring. Depths are to 20 feet in the well-lit shafts. The springs are about 50 feet apart.

The clean, modern fish camp provides many services, including a boat ramp, boat dockage, canoe and shallow-draft boat rentals, gas, bait, ice, groceries and more. They also operate a campground and canoe trail service.

Crystal River   Roy Stoutamire

NO.6 JOE'S SINK
Location: From US 19 at Weeki Wachee, turn east on S-50 and travel 2 miles. There are two sinks located 30 yards off the side of the highway and about 50 yards past the power lines. They are in an open area and can be seen easily from the highway.

The waters of Joe's Sink offer the coldest freshwater dive in the state. The temperature drops from the 70s on the surface to the low 60s as you descend to the bottom.

The sink has a peanut-shaped appearance on the surface, with the non-connecting sinks resting at each end. The west sink has a neck about ten feet in diameter which widens to 30 feet at the bottom, in 45 feet of water. The walls have a crater-like surface. Natural light vanishes at 30 feet in both sinks.

The east sink is similar in shape, but drops to a depth of 75 feet and is larger in diameter at the floor.

NO.7 WEEKI WACHEE RUN
Location: Turn west off US 19 onto S-50 at Weeki Wachee. Go 3.6 miles to S-595. Turn south and go 1.5 miles to the bridge (Rogers Park).

The park has a white sand beach and boat ramp. There are no boats for rent, so you must bring your own or catch a ride on one going upstream. Go to the private property line near Weeki Wachee and snorkel or tube back down the run. This very clear water runs for ten miles to the Gulf.

NO.7 HOSPITAL HOLE

Location: From Rogers Park on Weeki Wachee run, swim or take a boat upstream about 300 yards.

A large crevice of bone-white limestone opens on the south side of the run in about eight feet of water and drops to depths of 155 feet. From the 90-foot level on, there is a tea-colored layer of hydrogen sulfide gas which has an unpleasant odor and taste. Because of the depths and hydrogen sulfide layer, this must be considered an advanced cave dive.

NO.8 SAW GRASS SPRING

Location: From the junction of US 19 and S-50 in Weeki Wachee, go west on S-50 about 5 miles to an inn on the left. 150 yards past the inn, turn left onto a dirt road and travel .1 mile. Turn right on small dirt road and continue for 100 yards to the spring.

This is a beautiful little spring filled with waving saw grass. Depths are to 20 feet. There is saltwater in the spring, and visibility is usually 40 feet.

NO.9 PALM SINK

Location: From the junction of US 19 and S-52 south of Hudson, travel north on US 19 for 1 mile. Pull off to the left side of the road (west) just before the highway bears to the right. The sink is located just behind a highway guardrail.

Palm Sink has depths to 45 feet. An overhang on the south side can be penetrated only a short distance. This is an advanced cave dive.

NO.10 CRYSTAL SPRINGS

Location: Follow S-39 5 miles south of Zephyrhills to the town of Crystal Springs. The spring is located just west of town.

Several springs that form the headwaters of the Hillsborough River have been dammed to form a clear pool with depths from 13 to 15 feet. Sixty-five m.g.d. issue from the group of springs. The year around temperature is 74 degrees. The combination of clear water, aquatic plant life and numerous fish provides a great setting for the underwater photographer. A nice picnic area is located near the springs.

NO.11 LITHIA SPRING

Location: From Brandon, located 8 miles west of Tampa on S-60, turn south on S-674. In 6 miles, cross the Alafia River and continue .5 mile to springs.

Lithia is a very clear spring with shallow depths that offer excellent snorkeling around the boil area and down the clear run to the Alafia River. A nice picnic area is adjacent to the springs. No scuba is allowed.

5

SOUTH TALLAHASSEE

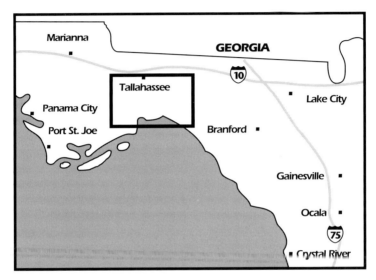

Spring Diving Locations

1. Gopher Hole Sink
2. Emerald Sink
3. Little Dismal Sink
4. Big Dismal Sink
5. Fish Sink
6. Upper River Sink
7. Promise and Go-Between Sinks
8. West Hole Sink
9. Wakulla Spring
10. Cherokee Sink
11. Wakulla River
12. Wacissa River Springs
    Wacissa River Clearwater
    Canoe and Snorkel Trail

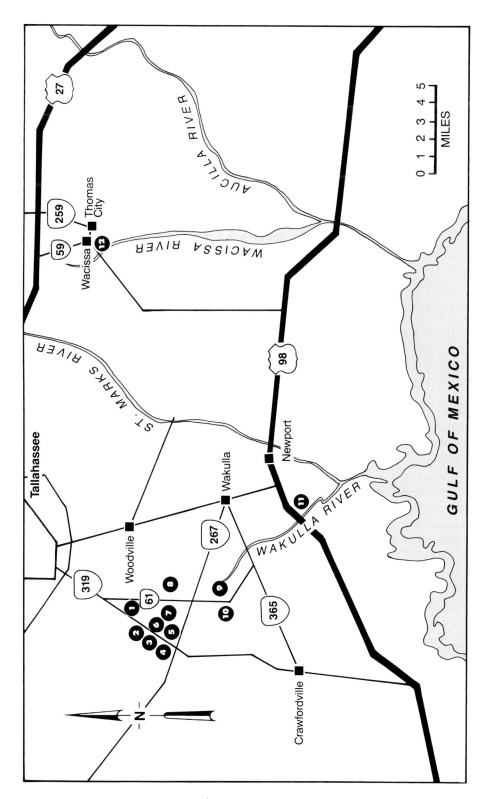

Area Information

A series of large, deep sinkholes surrounded by high steep walls typifies the diving locations found south of Tallahassee. Underwater exploration of these sinks is both interesting and exciting. Poor visibility and bad silting conditions are common, requiring extreme caution from all divers.

Located 14 miles south of Tallahassee is Wakulla Springs, one of the world's largest and deepest springs. It is now a state park, with glass-bottom boat trips. Scuba diving is not allowed. The diver will have to be satisfied to view the spectacle from a boat. Unfortunately, Natural Bridge Spring is now closed to divers.

Located east of Tallahassee is the beautiful spring-fed Wacissa River. This is a popular area for combining a canoeing-diving-camping trip. The 14-mile canoe trail winds through some of Florida's most beautiful countryside.

There are three diving retailers in Tallahassee to fill your tanks. Motels, restaurants and camping facilities are plentiful.

NO.1 GOPHER HOLE SINK

Location: From Four Points (a main intersection in south Tallahassee), travel south on US 319 for 6.6 miles (or .6 mile past C-260 intersection). Turn left off S-319 onto small dirt road. Bear right .1 mile to the sink.

Gopher Hole is a small sink with steep banks and depths to 100 feet. Two caves provide only a short penetration in either direction. This site is known locally as Oak Ridge Blue Sink.

NO.2 EMERALD SINK

Location: From Four Points (a main intersection in south Tallahassee), go south on S-319 toward Crawfordville for 8.2 miles. Turn right (west) on the first hard surface road past the Wakulla County sign (about a quarter mile). Look for the New Light Church sign at the turn. Go .3 mile to first sand road to the left. The sink is located less than 100 yards back and to the right

The basin of this beautiful clearwater sink is about 60 feet across. A wooden pier is at the water's edge for an easy entry. The sink's walls mushroom out as you drop below the surface. Maximum depth in the basin is 120 feet. A 15-foot wide cave opening narrows and goes to a depth of over 200 feet. Because of the depths,

Diving at night in a Florida spring.   Ned DeLoach

this is an extremely dangerous cave. Restrict your dive to the lovely basin unless you are a certified cave diver.

NO.3 LITTLE DISMAL SINK

Location: From Four Points, go south on US 319 for nearly 9 miles (or 1.3 miles past C-260 intersection). Turn right (west) onto dirt road. Wind to the right at the first fork and bear left at the second fork. The sink is .6 mile from the highway.

The area around both Little and Big Dismal Sinks is part of the Apalachicola National Forest. The Forest Service is making improvements around the sinks. A boardwalk, observation deck and platform for divers has been completed. Diving is allowed November until March. A large cave is located on the northwest wall. A 400-foot penetration can be made at a maximum depth of 60 feet.

NO.4 BIG DISMAL SINK

Location: Follow dirt road .2 mile west from Little Dismal Sink.

Big Dismal is the most recent sink formed in the Tallahassee area. It collapsed into the aquifer just after the turn of the century. The very large, impressive sink is surrounded by steep walls. An overhang is located 85 feet below the surface on the north wall. It can be penetrated for about 100 feet.

NO.5 FISH SINK

Location: From Four points follow the direction toward Emerald Sink. At the New Light Church sign turn left on C-319 (C.J. Spears Road). Make an immediate right

and go approximately 1 mile. Careful, this is 4-wheel drive territory. Look for a
rope swing hanging from a limb high over the water.

This is another spring-siphon with depths to 50 feet. The siphon cave is located on the south wall at a depth of 30 feet. The spring runs to the north. Because of the cave's depth, this must be considered an advanced cave dive.

NO.6 UPPER RIVER SINKS

Location: From Four Points, go southwest on US 319 for 10 miles. Turn left (east)
on first graded road past River Sink Grocery. Go .4 mile before turning left on small
dirt road (bad road, watch for deep ruts and loose sand). Go another .4 mile before
turning left once again. Continue for .8 mile, turning right at small dirt crossroad.
Sinks are 100 yards ahead.

This is a kidney-shaped sink with depths to 40 feet. A spring is on the north end and a siphon on the south. Lower River Sink is now on private property and inaccessible to the public.

NO.7 PROMISE AND GO-BETWEEN SINKS

Location: Follow directions to Fish Sink but look for depression to your right (east)
of dirt road. You can drive to Promise, but the going gets rough.

Promise is a shallow sink (depths to 30 feet) with a short 35-foot dive under a natural bridge to Go-Between Sink.

Go-Between is even more shallow than Promise (depths from 5 feet to 25 feet). The cave leads to a big room that starts 150 feet back. A foot trail running east leads to about 12 other small sinks. Nearly everyone snorkels these areas rather than lug tanks through the woods.

NO.8 WEST HOLE SINK

Location: From Four Points in Tallahassee, go south on US 319 to S-61. Two miles
south of Wakulla Co. line, turn left on sand road (.1 mile past white fence on right
side of S-61). Bear left for 300 yards to the sink. Enter water about midway on
west side of sink.

The sink depths vary from 45 to 80 feet. A 150-foot ledge with a 110-foot penetration is located on the sink's south side. The bottom is very silty and the limestone walls are much softer than they appear. Please do not disturb the many saltwater fossils in the limestone walls.

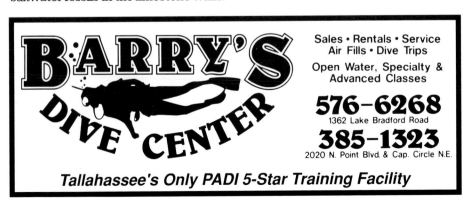

NO.9 WAKULLA SPRING

Location: 14 miles south of Tallahassee on S-61, turn left on the Wakulla Spring Road.

This is one of the largest and deepest springs in the world. In 1932 the site was opened as a resort and wildlife sanctuary. The state acquired the land in 1986. It is operated as a resort park with glass-bottomed boat trips over the tremendous boil and down the river. A huge cave opening, over 100 feet wide, begins at a depth of 125 feet and drops to 185 feet, discharging 183 m.g.d. that feeds the pristine Wakulla River. No scuba is allowed. Snorkeling is confined to a roped-off swimming area that does not extend over the cave opening. There are picnic tables and a large old hotel that is quite picturesque with a snack bar and dining room.

NO.10 CHEROKEE SINK

Location: Directly across S-61 from entrance to Wakulla Spring, follow the dirt road for .1 mile; then bear left for 1 mile to a small trail running off to the sink on the right.

Cherokee Sink is a large sink with thick woods surrounding the area. Depths go to 80 feet. There is little of interest, with visibility generally medium to poor, and a lot of trash in the water.

NO.11 WAKULLA RIVER

Location: This is a clear river fed by Wakulla Spring. Enter at the US 219 bridge.

There is wonderful snorkeling through eel grass and over the white sand bottom here. There is also good canoeing, with relics and fossils hidden under the sands.

NO.12 WACISSA RIVER SPRINGS

Location: From Tallahassee, travel east on US 27 about 20 miles. Turn right on S-59, continuing south for 4 miles to the S-259 intersection. Stay on S-59 for 1.7 miles to the headspring. When S-59 makes a sharp right turn, continue straight for 5 mile to the river.

OR

Exit I-10 on S-59 (Ext. 32) and travel 10.5 miles to the spring.

A beautiful, primitive spring-fed river flows south from to its junction with the Aucilla River. Big Spring, Garner Spring, Blue Spring, Buzzard Log Spring, Minnow Spring, Cassidy Spring and others form the head of the river. All these areas can be explored using a small boat or canoe. There are camping areas along the river, and the bottom provides excellent relic hunting. There is a sand boat ramp and swimming area at the headspring.

WACISSA RIVER CLEARWATER CANOE AND SNORKEL TRAIL

This lovely, wild Florida river is a true delight. The 14-mile canoe trail takes approximately 6 hours to complete. It starts at a large headspring pool (No.12) and flows south through swampy lowlands. Land elevation is seldom over three feet. The entire river is in the Aucilla Game Management Area. Goose Pasture is the only good spot for camping. It is nine miles downriver. Tables, grills, shelters and

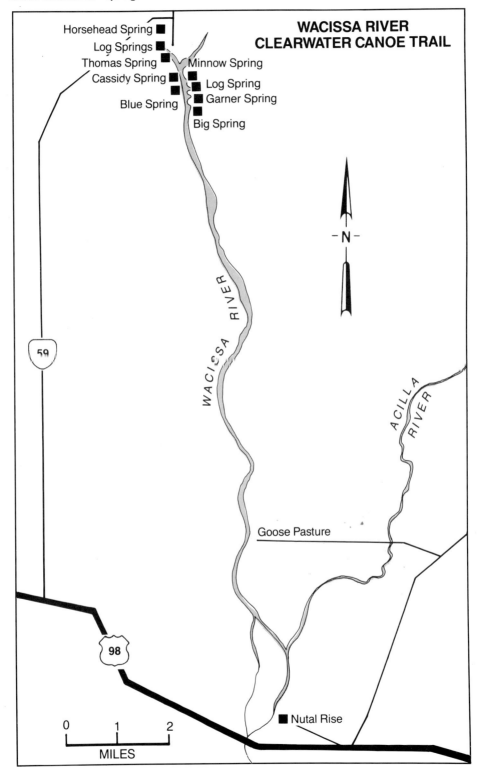

Horsehead Spring

Log Springs

Thomas Spring

Cassidy Spring

Blue Spring

Minnow Spring

Log Spring

Garner Spring

Big Spring

**WACISSA RIVER
CLEARWATER CANOE TRAIL**

– N –

WACISSA RIVER

ACILLA RIVER

59

Goose Pasture

98

0   1   2

MILES

Nutal Rise

sanitary facilities are available at this popular site.

There are 12 major springs that feed the Wacissa with clear water. They are all located within 1.5 miles from the headpool at the park. Horsehead Spring is .4 mile up narrow, shallow Horsehead Run. It is extremely difficult to reach this small spring. Log Springs is only .1 mile up the same run. The pool is about 40 feet at its widest, with depths to 25 feet at the two limestone vents that issue water into the basin.

Thomas Spring is in the main pool at the park. It is located about 100 yards northwest of the boat ramp. The 8-foot vent drops to a 28-foot depth.

Spring No. 1 is 75 yards southwest from the diving board. Depths are about 25 feet. Spring No. 2 is 15 feet south from the diving board. It is 16 feet deep.

Cassidy Spring is .5 mile downriver on the west bank. The large basin is 20 yards from the river. A wide run leads to the spring. Depths are 28 feet.

Minnow Spring is on the east side of the river, 150 yards south from Cassidy. Follow the side run 75 yards to the basin. Depths are only 8 feet on the north side over a vent 15 feet wide.

Blue Spring is 300 yards downstream from Cassidy. It is also on the east side. A round pool is 15 yards from the river. There is a maximum depth of 20 feet.

Buzzard Log Spring is directly across the river from Blue. Two vents drop to a depth of 8 feet.

Garner Spring is found on the west bank up a wide 50-yard-long run. There are two pools here, both shallow. This is one mile below the park.

Big Blue Spring has two wide runs to the Wacissa. Each is about 300 yards from the river. The large spring basin, 25 yards wide, has a maximum depth of 45 feet.

Divers and snorkelers often make a turn-around at Big Blue and canoe back to the park. Those who continue downstream will find the first high ground at Cedar Island, a mile below Big Blue and 2.5 miles from the park. The Wacissa Dam is 2.5 miles past the island. It was built to support an old railroad that crossed here years ago. It is 4 miles from this point to Goose Pasture. The first half of this section is narrow and swift in spots. Then the river widens once again and flows gently for about 2 miles to Goose Pasture.

To get to Goose Pasture by road, follow S-59 south for 14 miles to US-98. Go east on US-98 to the Aucilla River Bridge. After crossing the bridge, continue for 2 miles. Turn left on a wide graded road. Go 4 miles and turn left (west) at the intersection of another graded road. Continue for about 4 miles to the Goose Pasture Recreational Area.

It is 5 miles from the campgrounds to the boat ramp north of the US-98 bridge. About 2.5 miles past Goose Pasture there is a metal post to your right. This post marks the entrance to a canal dug by slaves in the 1850s. The canal's opening, which is often difficult to spot, is just to the post's right. The canal passes through a beautiful stand of cypress and oak to the Aucilla River. Canoe upriver .3 mile to the boat ramp.

6

PANHANDLE

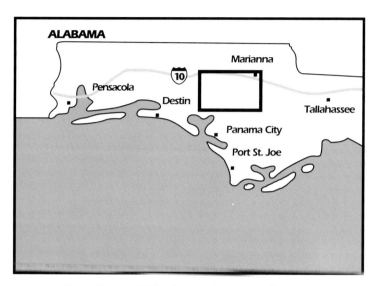

Spring Diving Locations

1. Merrit's Mill Pond/Blue Springs
   Twin Caves
   Shangri La Spring
   Indian Washtub
   Gator Spring
2. Blue Hole
3. Bozell Springs
4. Gadsden Spring
5. Cypress Spring
6. Becton Spring
7. Vortex Spring
8. Ponce de Leon Spring
9. Morrison Spring
   Chipola River Clearwater
   Canoe and Snorkel Trail

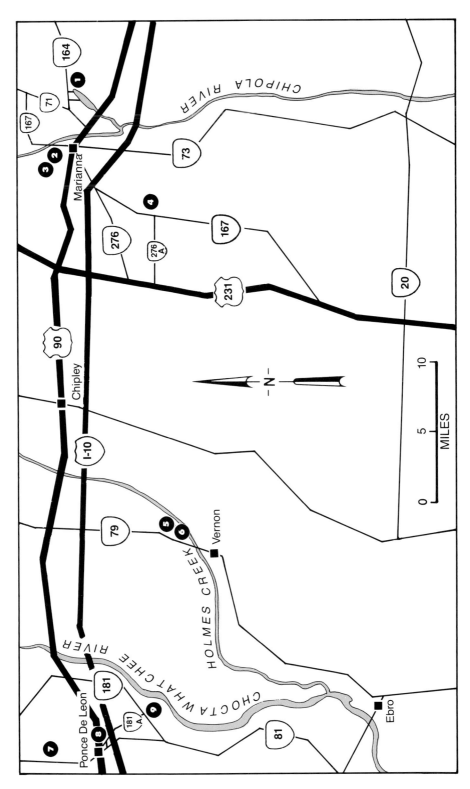

CHIPOLA RIVER

164

1

71

167

73

Marianna

2

3

4

276

276
A

167

231

20

90

Chipley

I-10

N

0   5   10
         MILES

79

5

6

Vernon

HOLMES CREEK

CHOCTAWHATCHEE RIVER

181

Ponce De Leon

181
A

9

8

7

81

Ebro

Area Information

Because of their close proximity, the springs located in the Florida Panhandle have long been popular with sport divers in the southern states. Morrison Spring, Vortex Spring, Cypress Spring and Merrit's Mill Pond are easily located and provide excellent diving year-around. Several locations are occasionally difficult to reach by car during the summer months because of poor roads and frequent rains.

Underwater relic hunting is at its best in the spring-fed Chipola River and Holmes Creek. Large, prehistoric shark's teeth and arrowheads are common discoveries for patient collectors. Air and divers' supplies can be found in Pensacola, Destin, Panama City, Vortex Spring, Morrison Spring and Dotham, Alabama.

NO.1 MERRIT'S MILL POND/BLUE SPRINGS
Location: From the courthouse square in Marianna, travel east on US 90 for 1.2 miles. Turn left on S-71 and go 1.1 miles and turn right onto S-164. After 3.5 miles, turn right at Blue Springs sign.

The pond is a large body of clear spring water that starts at Blue Springs pool and runs five miles to US 90. Both banks are lined with large cypress trees, making this a very picturesque locale.

Blue Springs is commercially operated from May through September. Admission is 25 cents. Picnic tables and a large bathhouse are located adjacent to the spring No scuba diving is allowed in the head pool, but snorkeling is permitted. Scuba is allowed in the rest of the pond area, which has an average depth of 15 feet. The bottom is covered with patches of aquatic plants, and fish are plentiful. Areas of Merrit's Mill Pond worth exploring are as follows:

NO.1 TWIN CAVES
Location: .5 mile downstream from Blue. You can get to the caves by boat from Hasty's Fish Camp, located .7 mile west on S-164 from Blue Springs Road (watch for the sign), or by snorkeling 250 yards east of Baptizing Landing, which is .5 mile west of the Blue Springs Road.

Two openings, 20 feet apart, connect with a room at 30 feet.

NO.1 SHANGRI LA SPRING
Location: Shangri La is located on the northwest side of the pond about 400 yards below Blue Spring, at the base of a 15-foot limestone cliff. Anchor in the pond, away from the private property adjacent to the area.

Very clear water boils from a shaft two feet in diameter at a depth of five feet. The cave opens into a room, 15 by 20, feet that slopes down to a depth of 25 feet. Water issues from a small slit at the deepest point. Shangri La offers good photographic possibilities if you are careful not to stir up the silt.

NO.1 INDIAN WASHTUB
Location: On the northwest side of the pond, directly from Twin Caves and just above Baptizing Landing.

This is an interesting area with depths to 20 feet. The water normally drains into tiny cracks in the bedrock and presumably exits at Hole-in-the-Wall Spring, nearly a mile away. During periods of heavy rainfall, the flow reverses, causing milky water to stream into the pond. This is a favorite spot for large black bass.

NO.1 GATOR SPRING
Location: Found on the southeast side of the pond, 3 miles below Washtub and directly across from the public boat ramp.

This is a beautiful, unusual and fun dive. The cave itself is somewhat of a rarity in Florida in that it is partially out of the water. You can snorkel more than 100 yards into the crystal clear cavern waters by carefully running a line through the air pockets. Be sure to bring lights because the cave makes a sharp bend to the right halfway back and all entrance light is lost. Depths vary from five to ten feet in the cavern stream. It is best to stay away from the floor because of its deep layer of silt. Large turtles are commonly sighted at the cave's entrance and, if you're lucky, you might spot an alligator.

The pond is dammed at US 90, forming a beautiful clear spring run that flows for about two miles to the Chipola River.

NO.2 BLUE HOLE
Location: Florida Caverns State Park, 3 miles north of Marianna on S-167.

Blue Hole is a large spring area with a run to the Chipola River. Snorkeling and scuba diving are allowed. A bathhouse and picnic area, as well as camping areas, are located near the spring. Be sure to take time to see the beautiful topside caverns while visiting the park.

NO.3 BOZELL SPRINGS
Location: Bozell actually consists of four separate clear springs in a line bisecting the Chipola River. To reach the springs, launch a boat at the ramp in Florida Caverns State Park and go about 1 mile upstream until you see the run from the main spring coming into the river on the right.

Spring I is a vertical fissure in a small bay on the west side of the river. It has a strong flow and depths are from 10 to 15 feet. Spring II is located at the bottom of the river and also has a strong outflow. Spring III, found in a bay on the east side, is more moderate in flow and is normally covered with duckweed (tiny green surface plants).

Spring IV is by far the most interesting. It is located at the head of a 200-yard run where a very small, silty cave opens at a depth of 15 feet. Avoid tying your boat up to private property surrounding the springs.

NO.4 GADSDEN SPRING
Location: From Marianna, travel 4 miles southwest on S-275. Turn south (left) on S-167 and go 2 miles to a Standard Station. Turn left on dirt road across from the station and go .4 mile and turn right on first dirt road. Stay on this road for .9 mile to fork. Bear left at fork and continue .5 mile to spring.

The spring pool is 25 yards in diameter. The surface water is not very clear due to the lack of flow, but visibility improves with descent. A large limestone cliff drops 50 feet to a large cave entrance.

NO.5 CYPRESS SPRING

Location: From I-10 take S-79 (Exit 17) south for seven miles. Turn left onto a sand road (look for sign) that winds .8 mile to the spring area.

OR

From the bridge north of Vernon go 2-1/2 miles north to a sand road on the right.

Cypress Spring is a commercially operated recreational area that allows both snorkeling and scuba. A crystal pool, surrounded by a primitive Florida forest, narrows before flowing .3 mile to Holmes Creek. The basin is 150 feet wide and 25 feet deep near the spring vent. A small, 6 x 10-foot cave entrance leads to a room 40 feet wide and 15 feet high that drops to a depth of 70 feet. At maximum penetration, a diver can easily see the light from the entrance. Over 90 m.g.d. flow through the restriction.

Although the basin is in a cypress lowland, excellent campsites are located on a small rise 100 yards away. There is a bathhouse and small concession nearby. Canoes can be rented for a 3 or 10-mile trip down Holmes Creek.

It is important that divers coming to Cypress from other freshwater areas check their gear for aquatic plants that might contaminate the basin. No pets or night diving are allowed.

NO.6 BECTON SPRING

Location: Downriver from Cypress Spring by canoe. The owner is planning to open a road to the spring and rent canoes.

Becton is a large spring, about 250 feet in diameter, with a flow of 32 m.g.d. and depths to 25 feet. The flow issues from small crevices between giant boulders. The cave is not large enough to enter. The long, shallow run is filled with plants and fish.

NO.7 VORTEX SPRING

Location: From I-10 take S-81 north (Exit 15) for 4.8 miles. Turn right at the sign.

Vortex Spring is a large, commercially operated diving park. The head pool is 200 feet across and 50 feet deep inside the cavern's mouth. At 15 feet two underwater platforms have been conveniently placed for instruction. There is an artificial air pocket between the two structures. A handrail leads 400 feet back into the cave to a depth of 115 feet. A steel grading blocks the passage at this point.

The basin is fed by 25 m.g.d. of clear 68-degree water. The run, Blue Creek, is four miles long and excellent for canoeing or snorkeling. Thirteen species of freshwater fish live in the spring. The 360-acre property also includes a complete dive shop, two lodges, campgrounds and a restaurant.

NO.8 PONCE DE LEON SPRING
Location: From the town of Ponce de Leon, travel south .3 mile on S-181A to the park entrance.

Ponce de Leon Spring is operated by the State. A bathhouse and picnic area are provided. The spring itself consists of a small head pool with depths to 15 feet. No scuba diving is allowed, but it is not needed to enjoy the shallow depths. An underwater natural bridge is formed by two cave entrances at 15 feet.

NO.9 MORRISON SPRING
Location: Travel south from the town of Ponce de Leon for 4.8 miles on S-181A. Turn left on the dirt road and continue 1 mile to the spring.

Morrison Spring is one of the finest freshwater dives in Florida. The huge spring basin and .5 mile run to the Choctawhatchee River are bordered by moss-covered cypress trees. The spring now has a diving concession with air, rentals and a snack bar. There is an entrance charge of $5.00 per day for divers. On-site camping is available for $2.50 per night.

The spring pool slopes to a large limestone cliff that drops off suddenly to 50 feet. There are two caves that offer excellent ledge diving. Neither can be penetrated beyond the glow of natural light. The first cave entrance is large and begins at a depth of 30 feet. The second cave entrance is much smaller and lies at a depth of 50 feet. It soon opens into a large room with depths to 90 feet.

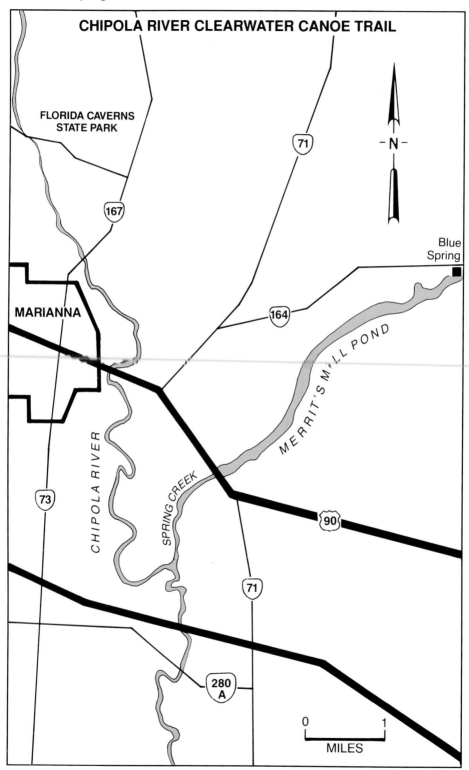

CHIPOLA RIVER CLEARWATER CANOE TRAIL

FLORIDA CAVERNS
STATE PARK

71

-N-

167

Blue
Spring

MARIANNA

164

MERRIT'S MILL POND

CHIPOLA RIVER

SPRING CREEK

73

90

71

280
A

0   1
MILES

CHIPOLA RIVER CLEARWATER CANOE AND SNORKEL TRAIL

The Chipola River is one of Florida's favorite underwater artifact hunting areas. Thousand of flint points and huge prehistoric shark's teeth have been found by divers. During low rainfall periods, the underwater visibility is good in many sections. This is due to the large volume of clear spring water that feeds into the river. There are several good accesses to the stream both above and below Marianna. Canoe trips lasting several hours or several days can be planned.

This wild, scenic river begins in south Alabama but doesn't form a navigable waterway until it flows into the Florida panhandle 18 miles above Marianna.

Bozell Springs (No.3) can be reached by canoeing one mile south from the Florida Caverns State Park, or by putting in on the southwest side of the bridge over S-162 and going with the flow south for 4 miles. This bridge is reached by traveling 7 miles north on S-71 from Marianna. Turn left (west) at Greenwood on S-162 and go a few miles to the river.

The Chipola goes underground at the State Park boat ramp and reappears .6 mile south. For a longer river trip put in at the S-167 bridge just north of Marianna. It is a nice 10-mile run from here to the S-280A bridge one mile south of I-10.

The first site of interest is Sand Bag Springs run. This is one mile from the bridge. The river then passes the eastern outskirts of Marianna until it reaches the US 90 bridge. This is not a good take-out point. One mile from US 90, at the point where an old railroad line crosses the river, a cave can be spotted in a high limestone bluff. On the west bank is a small spring-fed pool. Alamo Cave is .5 mile south at the end of the bluff. Dykes Spring run is just downstream on the east bank. It is 50 yards up the run to the spring. You can camp at this site.

The clearwater run from Spring Creek is six miles from Dykes Spring. A strong flow makes it difficult to canoe up this stream. One mile farther is the I-10 bridge. The take-out is less than a mile from here on the southwest side of the S-280A bridge.

Spring Creek is fed by the waters from Merrit's Mill Pond (No.1). Put in below the dam at the pond's south end where US-90 crosses. This is a beautiful 2-mile stream that flows through a swampy area to the Chipola River.

There is another 10-mile canoe trail below the S-280A bridge. To find the bridge turn south off I-10 and travel .4 mile on S-71. Turn right (west) on S-280A. It is less than a mile to the bridge. Several islands and shoals are passed on this section. Dry Creek enters from the east seven miles downstream. The take-out is at the boat ramp on the northwest side of S-278. This bridge is found six miles south from I-10 on S-71. Turn right (west) and go .4 mile to the river.

The Chipola continues as a navigable waterway for another 30 miles to Dead Lake. The take-out is just above the lake at the S-71 bridge. This is 15 miles south on S-71 from Blountstown.

7

PENSACOLA

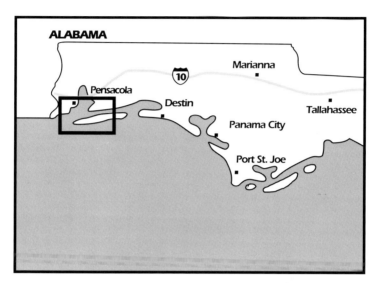

Offshore Diving Locations

1. Ft. Pickens Jetties
2. *Sport*
3. *Catherine*
4. Unknown Wreck
5. USS *Massachusetts*
6. Three Coal Barges
7. Casino Rubble
8. Air Transport
9. Liberty Ship
   (*Joseph L. Meek*)
10. *Tex Edwards* Barge
11. Bridge Rubble
12. Russian Freighter—
    San Pablo
13. Monsanto Boxes
14. P.C. Barge
15. *Sylvia*
16. *Deliverance*
17. *Tessie*
18. Three Deck Tug
19. Trysler Grounds
20. *Heron* & LCM
21. Railroad Bridge Rubble
22. Mr. Green's
23. A-7 Jet
24. Tenneco Oil Rig
25. Brass Wreck
26. *M.D. Whiteman*
27. Timberholes
28. Liberty Ship
    (*Joseph E. Brown*)

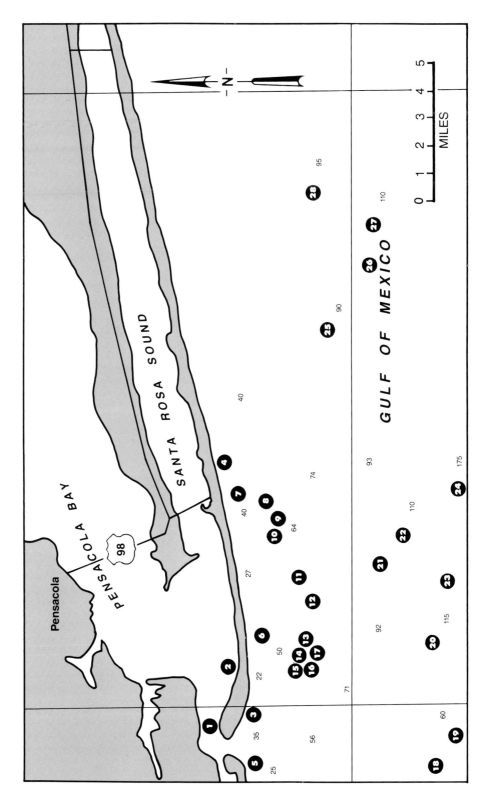

Pensacola

PENSACOLA BAY

98

SANTA ROSA SOUND

GULF OF MEXICO

N

0 1 2 3 4 5
MILES

Area Information

During the past several years the combined efforts of the Escambia County Marine Recreation Committee, the Department of Commerce and the U.S. Navy have greatly enhanced the recreational opportunities for sport divers in the Gulf waters just out from Pensacola Bay. Through their cooperative efforts, many exciting artificial reefs have been added to the area's abundant natural ledges and wrecks. Here, divers have the choice of visiting sunken battleships, Russian freighters, fighter planes, oil rigs and more. All are active fish havens that have attracted grouper, snapper, barracuda, flounder, amberjack, and brilliant multi-colored tropicals by the thousands.

There are a few good shore dives in the area. However, the shallows out from the beaches are generally disappointing for divers. Their large flat expanses of white silica sand run unbroken for miles. The "hot" thing to do is to take a diving charter from Pensacola. Several excellent boats are available throughout the year. It is best to make advance reservations during the popular summer holidays.

Summer is the in-season for diving in the Gulf. Calm seas are common during the months of April through October. Water temperatures stay around 80 degrees. Underwater visibility averages 60 feet, with days of 100-foot visibility common. Spearfishermen will find the hunting good. Grouper, snapper, amberjack, and large Warsaw grouper are the most plentiful game fish. Large fish move into the area during the winter. Underwater photography, tropical fish and shell collecting are good at all dive locations; and, of course, the wreck diver will be in paradise.

NO.1 FT. PICKENS JETTIES (Beach Dive)
Location: On the western tip of Santa Rosa Island National Seashore.

The Ft. Pickens jetties is an excellent beach dive with easy access. A drive approximately six miles west on Santa Rosa Island will bring you to the park. Here you are only minutes away from the hotels, restaurants and clubs. The park's excellent facilities include camping grounds. The rock jetties are located at the very end of the island. Start your dive at the beach and follow the gradual slope to a 50-foot depth. The rocky bottom is alive with marine life. The site is often used by instructors for check-out dives. Because of strong currents that accompany each tidal change it is extremely important to dive on a slack tide. Check with the local dive shops or the park rangers for tide information. Remember to always tow a diver's flag on the surface.

NO.2 *SPORT* (Beach Dive)
Location: West end of Santa Rosa Island on the bay side.

This wreck is located in the bay just past the entrance to the Gulf Islands National Seashore. It is accessible by beach or by boat. The *Sport* is an old tug that was sunk during the 1906 hurricane. It is located in shallow water on a sand bottom. Check with local dive shop or park ranger for the exact location.

NO.3 *CATHERINE* (Beach Dive)

Location: West end of Santa Rosa Island on the Gulf side, inside the Gulf Islands National Seashore. It is just off the beach near the Old Coast Guard Station.

The *Catherine* was a Norwegian bark that ran aground on August 7, 1894. The broken remains lie in approximately 15 feet of water. A dive from the beach will require a strong kick to make it through the surge. Remember to float a diver's flag behind.

NO.4 UNKNOWN WRECK (Beach Dive)

Location: Just off Pensacola Beach. Swim east toward the second water tower.

This old, unidentified wreck is just beginning to uncover itself. It lies just off the beach in 15 feet of water. Already uncovered are ballast stone, a steering station and a large metal tank.

NO.5 USS *MASSACHUSETTS*   13215.0   47108.9

Location: A little over a mile off the rock jetties, this wreck is found easily.

This is one of the best small boat dives in the Pensacola area. The 500-foot battleship of WWI vintage was sunk by the Navy in 1927. Lying in 25 feet of water, part of the ship is still exposed. Though it is mainly intact, some sections of the USS *Massachusetts* are covered by sand. In winter, diving can be hampered by rough surge.

NO.6 THREE COAL BARGES   13270.6   47107.6

Location: 1.8 miles off the beach, in 50 feet of water.

These three barges rest end to end on a white sand bottom, and form a wonderful area for safe, easy diving. The top decks of the 200-foot barges are 15 feet off the bottom. The area has developed into an outstanding fish habitat. The clean sand surrounding the ships is covered with large sand dollars and other shells.

NO.7 CASINO RUBBLE   13326.3   47116.0

Location: 1 mile off Pensacola Beach.

The rubble from an old casino (the first building constructed on Pensacola Beach) was dumped in 60 feet of water to form an artificial reef. Large concrete bricks and other construction materials provide habitat for flounder and red snapper.

NO.8 AIR TRANSPORT
Location: 7-1/2 miles east-southeast of the pass leading to Pensacola Bay.

The remains of a large 128-foot air transport were sunk five miles off Pensacola Beach in 75 feet of water. Its tail section rises 28 feet off the sand bottom. The sunken plane is just one more example of Pensacola's outstanding program to provide interesting diving for the public.

NO.9 LIBERTY SHIP (*JOSEPH L. MEEK*)   13306.7   47102.7
Location: 7 miles east-southeast of the pass leading to Pensacola Bay.

The intact hull of the 480-foot Liberty Ship *Joseph Meek* was sunk by the Department of Commerce in November 1976 as part of their program to form areas for sport divers and fishermen. She rests in 80 feet of water,with her sides rising 20 feet off the flat bottom.

NO.10 *TEX EDWARDS* BARGE   13300.3   47101.8
Location: 6-1/2 miles east-southeast of the pass leading to Pensacola Bay.

This large, intact deck barge is considered by charter boat captains to be one of Pensacola's safest dives. Blue angelfish and other tropicals hide in the many compartments. The top of the barge is at a depth of 60 feet.

NO.11 BRIDGE RUBBLE   13277.8   47091.9
Location: 7 miles from Pensacola Beach.

Twelve barge loads of rubble from the old Pensacola toll bridge were dumped in 75 feet of water to form an artificial reef. The large, complete bridge spans an area nearly 300 feet in diameter, forming an exceptional fish haven. Snapper, grouper and flounder are common at the site. The remains of a 100-foot barge lie at the western end of the area.

NO.12 RUSSIAN FREIGHTER—*SAN PABLO*   13263.6   47077.1
Location: 9 miles off Pensacola Beach

The *San Pablo* was torpedoed in the Florida Straits during WWII. She went down nine miles off Pensacola Beach while being towed to Mobile for repairs. She was later dynamited to clear shipping lanes. Her stern section and boilers remain intact in 75 feet of water. Her remains form an excellent fish habitat with many barracuda, grouper and snapper.

NO.13 MONSANTO BOXES   13248.6   47081.2
Location: 8 miles south-southeast of Pensacola Pass.

Over two hundred 4x4-foot fiberglass shipping containers with metal edges were welded together in units of eight or ten each and placed down in 70 feet of water as a fish haven. They have worked so successfully that the area is known by local divers as the Grouper Condos. A second Monsanto site is just southwest of the first. **13246.7   47079.7**

NO.14 P.C. BARGE   13253.6   47076.3
Location: 8 miles south-southeast of Pensacola Pass. Just east of the Sylvia.

A large barge was sunk as part of Escambia County's ongoing artificial reef building project. The barge is part of a cluster in this two-square-mile area that includes the Monsanto Boxes, the tugs *Sylvia, Deliverance* and *Tessie.*

NO.15 *SYLVIA*   13252.5   47075.5

Location: 8 miles south-southeast of Pensacola Pass.

This intact 65-foot tug rests on a sand bottom in 82 feet of water. There is a lot of fish activity around the vessel and her surrounding sands are littered with sand dollars, starfish and shells. An excellent dive.

Paul Humann

NO.16 *DELIVERANCE*   13247.7   47074.7

Location: 8 miles south-southeast of Pensacola Pass. Just south of the Sylvia.

This is a another intact 65-foot steel tug in the artificial reef site.

NO.17 *TESSIE*   13250.0   47078.5

Location: 8 miles south-southeast of Pensacola Pass.

A 40-foot cabin cruiser, with her superstructure removed and filled with four auto bodies, was sunk in 75 feet of water as an artificial reef. The wreck is surrounded by large concrete culverts. Flounder are common in the area.

NO.18 THREE DECK TUG   13151.0   47050.3

Location: 11 miles south of Pensacola Beach in 95 feet of water.

A tug boat, with three levels of decking, sits intact on a clean sand bed. Large fish are common. Big sand dollars litter the ocean floor around the wreck.

Diver exploring a Gulf wreck.   Steve Straatsma

NO.19 TRYSLER GROUNDS

Location: 10 miles south-southwest of Pensacola Pass.

Basket sponges, soft corals and tropicals are common on this large broken bottom area. It is 60 feet to the bottom where lobster and game fish reside. Sixty-foot visibility can be expected during the spring and summer.

NO.20 *HERON* & LCM   13253.0   47060.6

Location: 11 miles south-southeast of the Pensacola Pass.

A steel tug rests upside down inside a LCM landing craft. This is a sight to see!

NO.21 RAILROAD BRIDGE RUBBLE   13255.1   47065.9

Location: 11 miles southeast of Pensacola Pass.

This is one of the many one-mile-square artificial reef sites permitted and developed by the Escambia County Marine Recreation Committee as an ongoing marine habitat project. Five huge concrete railroad bridge sections were put down. New materials are added as they become available.

NO.22 MR. GREEN'S   13279.9   47061.6

Location: 12 miles south-southeast of Pensacola Pass.

This is a large circular limestone reef in 110 feet of water. It provides divers with a great spot for lobster hunting, spearfishing and picture-taking. Visibility is usually good, between 50 and 100 feet.

NO.23 A-7 JET
Location: 17 miles south of Pensacola Pass in 110 feet of water.

The A-7 Corsair jet was lost as a result of a cold catapult from the deck of the aircraft carrier USS *Lexington*. The intact plane rests upside-down on the bottom.

NO.24 TENNECO OIL RIG   13324.1 47014.1 & 13324.5 47012.7
Location: 22 miles south-southeast of Pensacola Pass.

Two massive 500-ton structures were submerged in 175 feet of water after a 275-mile barge journey from its original Gulf location. This gift from the Tenneco Oil Company is the first use of a complete platform as an artificial reef. The first Loran coordinates are for the tower section with the deck intact. The second is for the section that consists only of the leg structures, called jackets. Diving should be limited to the rig's upper section which begins 80 feet below the surface. Visibility in the area is 100 feet or more during the summer months. Large fish are abundant.

NO.25 BRASS WRECK   13365.7 47085.0
Location: 14 miles southeast of Pensacola Pass.

This is a classic shipwreck. The vessel's identity is not known. Locals call her the Brass Wreck because of the many large brass pins that stick out from her ribs like tree limbs. The 250-foot wooden hull schooner sits in 90 feet of water on a clean sand floor. Considering her size, she probably had four masts and weighed over one thousand tons. Adorning the wreck are two large anchors, a pile of chain and a four-by-eight windlass. The broken wreckage swarms with sea life. Flounder, snapper and grouper lurk in the large ballast pile, while barracudas and amberjack hover above.

NO.26 *M.D.WHITEMAN*   13417.4 47080.6
Location: 17 miles east-southeast of Pensacola Pass.

An intact 65-foot steel tug put down as an artificial reef.

NO.27 TIMBERHOLES   13456.6 47074.4
Location: 18 miles east-southeast of Pensacola Pass.

Timberholes is a natural limestone reef in 110 feet of water with ledges that rise in places to 12 feet off the bottom. Lobstering, spearfishing, and shell collecting are popular activities on this beautiful north Florida reef.

NO.28 LIBERTY SHIP (*JOSEPH E. BROWN*)   13544.2 47062.7
Location: 20 miles east-southeast of Pensacola Pass.

Like all other Liberty Ships in the Gulf that have been sunk to become artificial reefs, the *Joseph E. Brown* is void of her superstructure. Her huge 500-foot hull now rests on a flat sea floor in 95 feet of water.

8

DESTIN

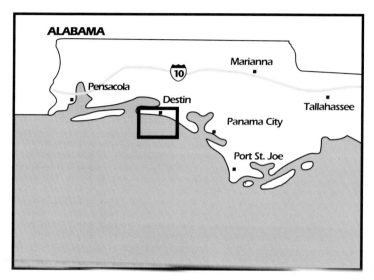

Offshore Diving Locations

1. Amberjack Rocks
2. Frangista Reef
3. White Hill Reef
4. Airplane Rocks
5. Liberty Ship (*Thomas Hayward*)
6. Destin Bridge Rubble
7. Air Force Barge
8. Brown's Barge

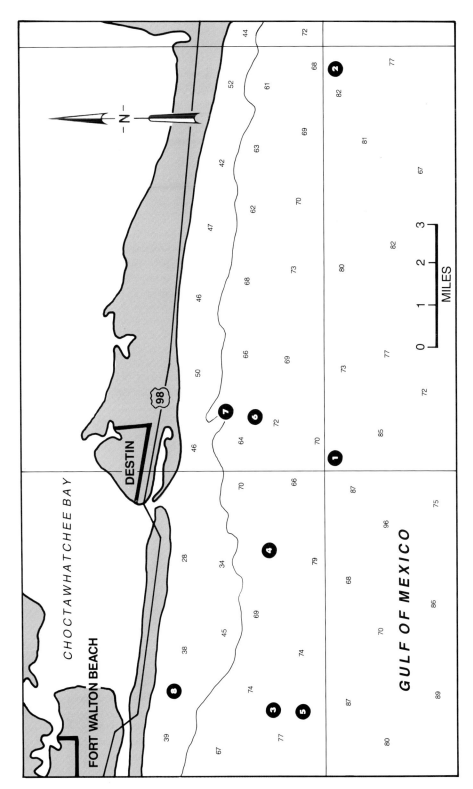

CHOCTAWHATCHEE BAY

FORT WALTON BEACH

DESTIN

98

GULF OF MEXICO

N

MILES
0 1 2 3

Area Information

Destin is located in the heart of one of America's most popular beach areas. Miles of beautiful white sand beaches extend both east and west. The area has been developed into a playground for tourists, with entertainment facilities everywhere along the beachfront. The water is clear and calm during the summer months, with underwater visibility ranging from 40 to 100 feet. The main feature of the underwater terrain off Destin is the prominence of limestone ledge reefs. These formations are part of the DeSoto Canyon that brings deep water closer to land here than at any other point in the Florida Gulf. Years of Air Force operations have caused many planes and missile parts to end up on the seabed, creating exciting underwater exploration. Spearfishing is popular and is allowed everywhere except near bridges, jetties and piers.

NO.1 AMBERJACK ROCKS
Location: 3 miles south of Destin Pass.

This is the largest and most popular reef in the area. Depths vary from 75 to 85 feet. The crescent-shaped rock reef is about 200 feet long with ledges 10 to 12 feet off the bottom. There are several ledge reefs in the area; many are undercut with small caves. Large cuts divide the reef sections. Basket sponges are one of the area's trademarks.

NO.2 FRANGISTA REEF
Location: 10 miles east of the Destin Pass.

A long five-foot-high ledge parallels the coast three miles out from the Frangista Hotel. This is a great spot to find grouper. Depths are 85 to 90 feet.

NO.3 WHITE HILL REEF   13659.8   47119.3
Location: 8 miles south-southwest from Destin Pass.

Depths range from 85 to 90 feet on this rocky reef area. Spearfishing and lobstering are excellent.

NO.4 AIRPLANE ROCKS
Location: 2 miles southwest of Destin Pass.

A popular rock reef area with depths to 70 feet. The site is directly below the approach to Eglin AFB.

NO.5 LIBERTY SHIP (*Thomas Hayward*)   13648.1   47115.7
Location: 6 miles southwest of Destin Pass.

The Liberty Ship *Thomas Hayward* was sunk by the Department of Commerce in 1977. She rests in 90 feet of water.

NO.6 DESTIN BRIDGE RUBBLE   13720.4   47131.0
Location: 1-1/2 miles southeast of Destin Pass.

In the mid-70s, the concrete spans from the old bay bridge were placed down to become an artificial reef. The rubble has attracted huge schools of baitfish which

bring in large schools of spadefish to feed. Depths are 60 to 65 feet. Check-out dives are frequently made here.

NO.7 AIR FORCE BARGE   13720.9   47132.8
Location: Just over a mile southeast of the Destin Pass.

This 100-foot barge, also known as the Elgin Barge, rests at 65 feet on a sand plain. Its deck is 55 feet down. Two large holes were blown in the side to sink the vessel. There are many schooling fish on the site and the sand around the wreck is a good area to gather large sand dollars. A nice spot for check-out dives.

NO.8 BROWN'S BARGE   13660.7   47134.1
Location: 3-1/2 miles west of Destin Pass.

A 200-foot barge and assorted rubble, including fast deteriorating boxcars. Depths are to 70 feet.

9

PANAMA CITY

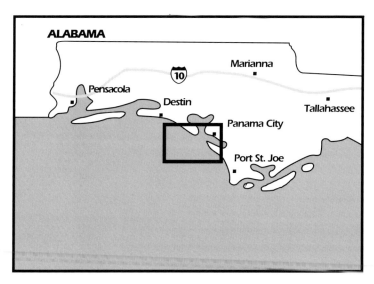

Offshore Diving Locations

1. Life Boats
2. Spanish Shanty Barge
3. St. Andrew's Jetties
4. *Simpson*
5. Navy Scrap Pile
6. LOSS Pontoon
7. *Chicksaw*
8. Stage I and Stage II
9. Three Airplanes
10. USS *Strength*
11. Midway Artificial Reef
12. Liberty Ship

13. Warsaw Hole
14. Quonset Hut
15. *Commander*
16. *Tarpon*
17. Fontainebleau Artificial Reef
18. Eleven Barges
19. Hathaway Bridge
20. Phillips Inlet
21. *Chippewa*

NOT ON MAP:
   Grey Ghost

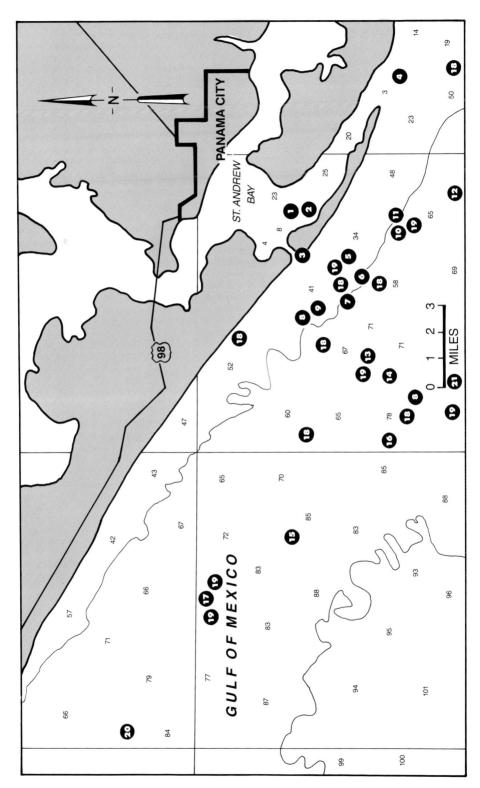

Area Information

Panama City is one of America's premier underwater recreational areas. Its north Florida location provides more diving and less driving for travelers from almost any part of the country.

In the mid-70s the Panama City Marine Institute's Artificial Reef Program started a reef-building tradition that has become a popular community project. Great ships, towers, bridge spans, airplanes and Naval scrap materials have been carefully placed on the ocean's floor forming prolific marine ecosystems. These exciting new dive sites complement Panama City's natural reefs and historic shipwrecks to create a variety of diving adventure that will keep you coming back year after year.

Note: While in Panama City, visit the Museum of Man in the Sea located on the Back Beach Road, 1/4 mile west of Highway 79.

NO.1 LIFE BOATS   14101.7   46987.4

Location: Inside St. Andrews Bay, about 1 mile north of Shell Island.

This is an interesting and safe dive site, consisting of five steel-hulled lifeboats which were cabled together and sunk in 25 feet of water. Five-to-15-foot visibility can be expected. Scattered marine life hovers about the area.

NO.2 SPANISH SHANTY BARGE   14101.4   46985.1

Location: Inside St. Andrews Bay, near the Spanish Shanty cove.

An old tar barge lies in 18 feet of water and is often used for check-out dives. Visibility ranges from 5 to 15 feet. Sheepshead and mullet are often sighted around the barge. Many shells are found near the 150-foot barge.

NO.3 ST. ANDREWS JETTIES (Beach Dive)

Location: At the west end of Shell Island in the St. Andrews State Park.

The depths and visibility off the jetties vary from 3 to 50 feet. Try to dive this site near a slack tide; tidal currents can be strong. During the summer, tropicals are everywhere. Spearfishing and marine specimen collecting are illegal here.

NO.4 *SIMPSON*   14121.6   46942.6

Location: Just outside the old pass, approximately 1/2 mile offshore.

The tug, *E.E. Simpson*, sank during rough weather in 1929 while trying to free a grounded fishing boat. It rests in 20 feet of water. Her smokestack can be seen just below the water on calm days. Because it is close to the pass, it is best dived at high tide.

NO.5 NAVY SCRAP PILE   14079.4   46973.4

Location: 3 miles off Shell Island.

This large area has been used for 20 years by the Navy as a artificial reef site. Chains, metal containers and old airplane parts are among the debris that can be found scattered about the bottom. The area has become a fish haven and offers good spearfishing, especially during the winter months when large fish come

HYDROSPACE DIVE SHOP

Gulf Coast's Largest & Most Experienced Dive Service

North Florida's Best Diving

1-800-874-3483

Daily Wreck & Reef Dives

Fast 48' Customized Dive Boats

Dive The Worlds Most Beautiful Beaches

6422 W. Hwy. 98
Panama City Bch., FL 32407
(904) 234-3063

3605 A. Thomas Dr.
Panama City Bch..FL 32407
(904) 234-9463

into the area. Flounder are also found around the reef. Four 20-foot destroyer-mooring buoys were sunk in late 1988, and a large barge was added in 1990.

NO.6 LOSS PONTOON   14078.0   46973.7
Location: 3 miles off Shell Island.

The LOSS Pontoon is a large metal cylinder which served as the working prototype for a Navy lifting pontoon. It is 15 feet in diameter, 40 feet long, and weighs just over 80 tons. It was designed to lift disabled or sunken submarines. The cylinder has been on the bottom at 60 feet since 1978, attracting all types of tropical and game fish.

NO.7 *CHICKASAW*   14056.8   46978.6
Location: 3 miles out, just seaward from the Panama City Sea Buoy.

The *Chickasaw* was a steel-hulled tug built at Pensacola in 1908. The 107-foot vessel was used to build the St. Andrews Jetties and perform other duties for the Army Corp of Engineers. She was retired after WWII. Later she served under the name *Sherman* VI until, nearly derelict, she sank in the late 70s. The vessel now rests in 75 feet of water, attracting sea life.

NO.8 STAGE I—STAGE II   13980.2   46957.9   &   14069.3   46997.8
Location: Stage I is 13 miles offshore; Stage II is 3 miles out.

These Navy research towers were similar to the oil rigs seen off the Louisiana Coast. In the summer of 1984, the top structure was dismantled and Navy divers blew the legs with explosives, allowing the supports and crossbeams to fall below the surface. The tangled structure of Stage I rises over 60 feet off the 107-foot bottom. Stage II has a 35-foot profile in 60 feet of water. The crisscrossing steel beams provide an excellent habitat for all types of Gulf fish.

NO.9 THREE AIRPLANES
Location: The first is just south of Stage II **14069.4   46997.1***, the second is near the P.C. Barge* **14042.5   47002.4***, and the third is on the Fontainebleau Artificial Reef Site* **14019.7   47028.5.**

A Navy T-33 trainer sits upside-down in 60 feet of water near the remains of Stage II. The other two are Air Force 101 Drones. Both are in 75 feet of water.

NO.10 USS *STRENGTH*   14076.7   46943.9
Location: 5-1/2 miles south-southwest from the St. Andrews jetties.

The *Strength* was a tender to a Navy minesweeping fleet until the mid-60s when it was retired and given to the Navy's Salvage Diver School. In 1987, she was sunk by the Experimental Dive Unit to test new explosives.

The vessel is 184 feet long with a 33-foot beam. She rests in 72 feet of water, with her deck only 40 feet below the surface. This is one of the area's most popular dives.

NO.11 MIDWAY ARTIFICIAL REEF   14072.6   46949.6
Location: 4-1/2 miles southeast of the St. Andrews jetties.

The Midway Site is halfway between the jetties and the Liberty Ship. The 3,000-yard-long permitted site is 72 feet deep. A lot of low-profile material is scattered around the area. The big draw is the body of Sikorsky 76 air-rescue helicopter and the prototype of the LOSS Pontoon. In 1988, two Bayline railroad cars were sunk here. Each boxcar is 60 feet long, 9 feet wide and 11 feet high. Large holes were cut in the sides and the doors were left ajar. Unfortunately, the boxcars are deteriorating rapidly.

NO.12 LIBERTY SHIP   14064.9   46918.7
Location: 7-1/2 miles south-southwest of the St. Andrews jetties.

   The Florida Department of Natural Resources sank a 441-foot Liberty Ship hull in July 1977. This huge steel hull is 57 feet across and rises 20 feet from a depth of 72 feet. Although cut at her waterline, the vessel still has several bulkheads in place. While diving here be on the lookout for the sea turtles and giant manta rays that frequent the site during the summer.

NO.13 WARSAW HOLE   14033.8   46969.5
Location: 9 miles southwest of the St. Andrews jetties.

   This is the closest of Panama City's many limestone reefs. It is horseshoe-shaped with a maximum depth of 85 feet. The ledge is 5 to 6 feet high and is covered with soft coral and sponges common to the local reefs. Game fish move in and out, but thousands of lovely tropicals and crustaceans have their permanent residence here.

NO.14 QUONSET HUT   14011.1   46966.9
Location: Approximately 10 miles southwest of the St. Andrews jetties.

   A 44-foot long, 30-foot diameter cylinder once used for Navy training exercises

was abandoned in 87 feet of water. It is now a spearfishing hot spot. The amberjack and barracuda of the summer give way to grouper and snapper in the winter. Visibility is generally around 25 feet, but rises to 80 feet at times.

NO.15 *COMMANDER*   13968.0   46982.3
Location: 15 miles west of the St. Andrews jetties.

This 65-foot tug wreck was discovered by two area captains in 1979. She sits intact and upright in 96 feet of water. This is a beautiful wreck that attracts many game fish.

NO.16 *TARPON*   13979.5   47001.7
Location: 9-1/2 miles west of the St. Andrews jetties.

The *Tarpon* was a 160-foot coastal freighter that served the coast between Mobile and Apalachicola. Eighteen of her crew members lost their lives in a 1937 storm that sank the vessel. She now rests in 92 feet of water. Her smokestack rises above the deck, and part of the stern and bow are still intact. Scattered around the site are thousands of 1937 vintage bottles—what remains of her lost cargo of beer.

NO.17 FONTAINEBLEAU ARTIFICIAL REEF   14019.7   47028.5
Location: 9 miles west of the St. Andrews jetties.

This site was one of the first artificial reef areas promoted by the Panama City Marine Institute. Material on the 72-foot bottom includes low-profile concrete and steel structures, an Air Force Drone jet, plus two 60-foot boxcars.

NO.18 ELEVEN BARGES

Location: Scattered (see the area map).

Barges were the first and most common large objects used as artificial reefs. Each is host to a full complement of Gulf sea life. A few have holes that allow penetration by advanced divers. The barges are listed below with their depths.

Blown-up Barge	65 feet	14052.4	46992.5
Deep Barge	92 feet	13979.8	46962.9
Smith Barge	70 feet	14066.9	46976.0
PCMI Barge	72 feet	14042.8	46999.8
Inshore Twin	71 feet	14069.0	46968.0
Davis Barge	55 feet	14116.0	46916.4
Holland Barge	65 feet	14065.7	46981.1
Long Beach Barge	50 feet	14067.3	47018.3
Spanish Shanty	18 feet	14101.4	46985.1
Offshore Twin	73 feet	14067.7	46967.0
B &B Barge	42 feet	14087.3	46970.7

NO.19 HATHAWAY BRIDGE

Location: 19 sites (see the area map)

The old Hathaway Bridge connected Panama City and the beaches until the new bridge was put in service in the early 60s. Until 1988, the old structure sat in place rusting away. The Florida Department of Transportation funded the project to remove the bridge and place it offshore as an artificial reef. The Panama City Marine Institute coordinated the spotting and permitting of the 19 new reef areas for divers and fishermen. The shallowest site is 42 feet and only two miles from the jetties; the deepest is 125 feet and approximately 15 miles out. The large spans were placed as single units. Their size, 165 feet long, 22 feet wide with a vertical profile of 36 feet, insures their success as fish attracters. The smaller deck spans were placed in groups of three and four. Each deck is 49 feet long by 22 feet wide and has a 6-foot profile. The project was completed in May 1988.

#1	14070.6	46953.1		#11	14003.8	46790.3
#2	14068.8	46949.0		#12	14074.5	46946.3
#3	14002.4	46914.3		#13	13995.2	46923.3
#4	13997.9	46915.8		#14	14037.2	46977.4
#5	14019.3	47031.1		#15	14025.2	47030.3
#6	14020.1	47022.8		#16	14031.8	46977.0
#7	13949.8	46950.0		#17	14112.6	46841.1
#8	13953.8	46955.8		#18	14002.1	46910.2
#9	13955.4	46961.0		#19	14085.4	46974.0
#10	13952.7	46969.6				

NO.20 PHILLIPS INLET

Location: 13 miles west of Panama City.

This area is virtually covered with reefs, ranging in depth from 60 to 110 feet. Sandstone and coral formations provide excellent spearfishing. The inlet is the

Chippewa

Danny Grizzard

stopping ground for transient Florida spiny lobster in the six- to 12-pound range. The reefs vary in size from two to four feet and range in length up to a half mile.

NO.21 *CHIPPEWA*   14012.2   46921.0
Location: 11.5 miles south-southwest of the St. Andrews jetties.

In February 1990, the Navy Experimental Dive Unit based in Panama City, sank the 205-foot working tug *Chippewa* in 96 feet of water for divers to enjoy. What a wreck! She sits upright and beautiful on a white sand plain. Unlike most vessels that are first stripped of deck machinery before being sunk, the *Chippewa* has her full complement of davits, wenches, levers, stairs and companionways intact. The deck is 70 feet below the surface. The large, open cabin is just 50 feet down.

The tug was launched in 1942 and worked in the Caribbean, using Trinidad as home base. In 1947, she was decommissioned and placed in reserves until 1989 when assigned to the the Experimental Dive Unit for use as a training project in salvage and ordnance techniques. This wreck alone is worth a trip to Panama City.

GREY GHOST (Not on map)   13891.11 46991.7
Location: 22 miles west of the St. Andrews jetties

The Panama City Marine Institute's Artificial Reef Program began July 12, 1978, with the sinking of this 105-foot military tugboat. The Ghost is lying on her port side in 108 feet of water near a limestone reef. The steel hull and superstructure are intact. This is a dramatic wreck not to be missed.

10

PORT ST. JOE

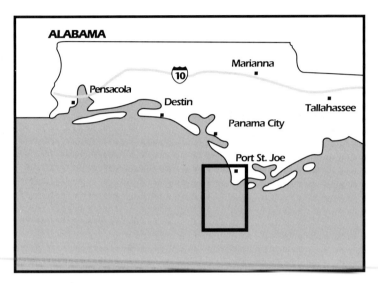

Offshore Diving Locations

1. Sharks Hole
2. Lumber Ship
3. *Kaiser*
4. Box Cars-Port St. Joe
5. Bridge Rubble-Port St. Joe
6. Sunken Barge
7. Bill's Barge
8. Hathaway Bridge-
   Port St. Joe
9. Shrimper's Junkpile
10. Eddy's
11. J.C. Reef
12. Gatewood Barge
13. Catherine's Kitchen
14. *Capt Jim*
15. *Capt Kato*
16. Barrier Dunes Barge
17. LST-Port St. Joe
18. Cape San Blas Lighthouse
19. *Birmingham Queen*
20. Schooner-Port St. Joe
21. *Audrey*
22. Cape Buoy
23. *Empire Mica*

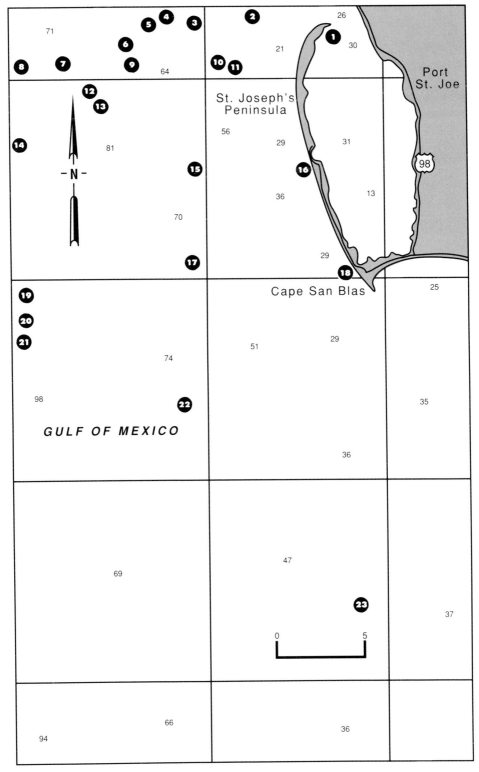

71

5 4 3

6

8 7 9

64

2

26

1

21

30

10 11

St. Joseph's
Peninsula

Port
St. Joe

12

13

56

14

81

29

31

98

- N -

15

16

13

36

70

36

17

29

18

25

Cape San Blas

19

20

51

29

21

74

98

22

35

GULF OF MEXICO

36

69

47

23

37

0   5

25

37

66

36

94

Area Information

Excellent Panhandle diving continues in the undiscovered coastal area west and south of the St. Joseph Peninsula. The St. Joe Bay serves as a prolific nursery for fish and crustaceans that spill out into the surrounding Gulf waters. Historic shipwrecks and recently constructed artificial reef sites boil with sea life.

Depths vary from a few feet in the Bay for beginning divers and snorkelers to over a hundred feet only seven miles offshore. The area's premier dive is on the broken remains of the Empire Mica, one of Florida's most fabled wrecks. The site is only an hour's run south from Port St. Joe.

NO.1 SHARKS HOLE (Beach Dive)

Location: Inside the Bay, just off the beach near the northern tip of the St. Joseph Peninsula. Must go by boat; there are no roads leading to the area.

A shell wall has been formed by the tidal currents that sweep out of St. Joseph Bay. Miles of beautiful open beach stretch south. The bank is in 10 feet of water and varies in height from three to 30 feet.

NO.2 LUMBER SHIP   14142.9   46830.5

Location: Two nautical miles from shore; just north of the entrance channel.

The scattered remains of this old 170-foot lumber hauler are in 22 feet of water. The wreckage, including the rudder post, and anchor chain, rises 10 feet off the sand. The shallow depth and its close location to the cut make this a nice spot for a night dive. Watch for octopus moving around the wreckage after dark.

NO.3 *KAISER*   14121.4   46840.6

Location: Approximately 1.5 nautical miles northwest of the sea buoy.

An old wood and steel tug sank in 1932. The hull has broken down, but the boiler and heavy deck machinery remain. The site has an eight-foot profile off a 45-foot sand bottom. Plenty of blue angelfish, planehead triggerfish and occasional turtles about.

NO.4 BOX CARS-PORT ST. JOE   14116.6   46845.5

Location: Approximately 3 nautical miles northwest of the sea buoy.

This is an old artificial reef site that has been acquiring a variety of material for 20 years. The collapsed frames of two railroad boxcars are surrounded by car bodies and tires.

NO.5 BRIDGE RUBBLE-PORT ST. JOE   14112.6   46841.1

Location: Less than 1/2 mile southwest of the Box Cars.

Concrete and steel debris provide a home for baitfish and tropicals in 60 feet of water. Occasional grouper and mackerel are sighted.

NO.6 SUNKEN BARGE   14098.1   46840.9

Location: 3.5 nautical miles west of the sea buoy.

This old barge is a good spot for spearfishing. Gag grouper and, occasionally, snapper frequent the site. Bottom depths are 70 feet.

NO.7 BILL'S BARGE
14059.1   46836.8

Location: 6 nautical miles west of the sea buoy.

An old barge that flipped upside down when it sank rests in 80 feet of water. The center has collapsed, but the four corners rise six feet above the sand. This is a good fish holder, with snapper, grouper and flounder about.

NO. 8 HATHAWAY BRIDGE-PORT ST. JOE
14003.8   46790.3

Location: 10 nautical miles west of the sea buoy.

A section of the old Hathaway Bridge from Panama City was sunk in the Port St. Joe waters. It rests at 100 feet. The span is 165 feet long and rises 36 feet off the bottom.

NO.9 SHRIMPER'S JUNKPILE
14104.0   46815.0

Location: 2 nautical miles southwest of the sea buoy.

A few years back a group of Carolina shrimpers worked the area. Not being familiar with the Gulf bottom, they brought up in their nets an unusual assortment of material placed down by private fishermen to attract fish. It was collectively dumped on this site. Expect to find car bodies, refrigerators, stoves, boats and plenty of fish. Depths are 72 feet.

NO.10 EDDY'S
14115.8   46804.3

Location: 2 miles south of the sea buoy.

The city of St. Joe and the State funded this artificial reef site and began to put

down material in 1989. The reef covers a hundred square yard area in 42 feet of water. Five tons of material, including 250 dumpsters, have a 16-foot profile in places. Don't overlook this spot; it is an excellent dive.

NO.11 J.C. REEF   14115.7   46804.1
Location: Just 100 yards east of Eddy's, inside the same permitted artificial area.

This site has been added to continually since 1968. A large number of concrete pilings rise two to three feet off the bottom.

NO.12 GATEWOOD BARGE
Location: 6 nautical miles southwest of the sea buoy.

A 70-foot barge in 80 feet of water is a private dive site maintained by Captain Black's Marine. No spearfishing is allowed. The steel sides are covered with invertebrates, offering some of the best macro-photography in the Gulf.

NO.13 CATHERINE'S KITCHEN
Location: Approximately 1/2 mile southwest of the Gatewood.

An artificial reef in 85 feet of water made up of box cars, tires and steel. A fish heaven for sheepshead, barracuda and grouper.

NO. 14 *CAPT JIM*
Location: 22 nautical miles west-southwest of the sea buoy.

A 65-foot shrimp boat sits upright and intact with her outriggers extended in 110 feet of water. The steel vessel rises 30 feet off the bottom.

NO.15 *CAPT KATO*
Location: 5 nautical miles west of Cape San Blas.

A 65-foot shrimp boat burned to the water line and sank in 65 feet of water in 1988. Her outriggers are still out. The remains of the fiberglass hull have become a great fish attracter for grouper and snapper. There are many large stingrays in the area.

NO.16 BARRIER DUNES BARGE   14114.9   46741.6
Location: 1 nautical mile southwest of Eagle Harbor on the St. Johns Peninsula.

In 1987, a barge loaded with heavy machinery started taking on water in high seas. The tug captain headed for the beach, but the 170-foot barge sank short of the mark. The equipment was salvaged, but the intacy barge remains in 30 feet of water. She is hard to enter; only three manhole covers are open. Currents have blown out a hole next to her side, uncovering remnants of an ancient forest. The barge now serves as a nursery for small snapper and cobia.

NO.17 LST-PORT ST. JOE   14061.9   46719.0
Location: 11 nautical miles south of sea buoy.

This old military landing craft sank during delivery in 1963 while rounding the Cape. The vessel, 75 feet long and 20 feet high, rises 10 feet from a 90-foot sand bottom. There is good visibility at this site. Often snapper are in the area.

NO.18 CAPE SAN BLAS LIGHTHOUSE 14113.2 46679.1

Location: 1/4 mile off Cape San Blas.

The remains of the old 65-foot #3 lighthouse is submerged in 25 feet of water. The bricks nearly rise to the surface. The structure, built in 1859, was the last brick lighthouse built on the site. Its light is still in use in the steel tower on shore.

NO.19 *BIRMINGHAM QUEEN* 13854.7 46735.3

Location: 22 nautical miles west of Cape San Blas.

An old steel party boat from Panama City has been down for many years. Only her sides stand. They rise eight feet from the 150-foot depths. For highly experienced divers only.

NO.20 SCHOONER-PORT ST. JOE 13902.0 46731.1

Location: 20 nautical miles west of Cape San Blas.

The broken ribs and anchor are all that remain of this wood and steel vessel that went down in the 1930s.

NO.21 *AUDREY* 13916.1 46721.1

Location: Over 20 nautical miles west of Cape San Blas.

Steel plates, 15 feet tall, are all that remain of this wrecked tug. The vessel rests in 100 feet of water.

NO. 22 CAPE BUOY

Location: 8 nautical miles off Cape San Blas.

Outer marker forms three miles of rock ledges. The rock is covered with soft corals and sea fans. Great area for clams.

NO.23 *EMPIRE MICA* 14023.5 46489.6

Location: 20 nautical miles south of Cape San Blas.

The *Empire Mica* is one of the most exciting wreck dives in Florida. The 465-foot British tanker was torpedoed by a U-boat on June 3, 1942. She drifted for 24 hours before sinking in 110 feet of water. One section of her deck is still upright and supports the wreck's symbol, an 18-foot spare propeller. Two torpedo holes can be found in her starboard side. Visibility is usually excellent, providing the underwater photographer dramatic images. Amberjack, snapper, barracuda and grouper are abundant.

11
ST. PETERSBURG
CLEARWATER – TAMPA

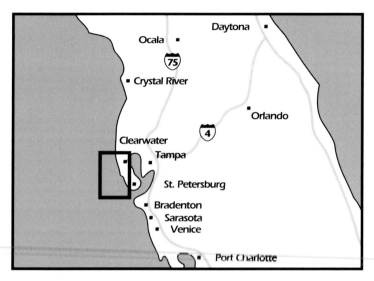

Offshore Diving Locations

1. Big Jack Hole
2. *Gunsmoke*
3. South Jack Wreck
4. Doc's Barge
5. South Jack Ledge
6. St. Petersburg Beach Reef
7. Twelve-Foot Ledge
8. Ten Fathom Wreck/
   Tramp Steamer
9. *Mexican Pride*
10. Treasure Island Reef
11. Mecco's Barge/*Betty Rose*
12. Indian Shores Reef
13. Pinellas #I/Rube Allyn
    Artificial Reef
14. *Blackthorn*
15. Tug *Sheridan*
16. Table Top
17. The Caves
18. Clearwater Wreck
19. 'G' Marker
20. Clearwater Reef
21. Dunedin Reef
22. Bomber
23. Masthead Ledge
24. Tarpon Springs Reef
25. Tugboat and Barge
26. Hellcat
27. Pasco County
    Artificial Reef 'PS'
28. Pasco County
    Artificial Reef 'CO'

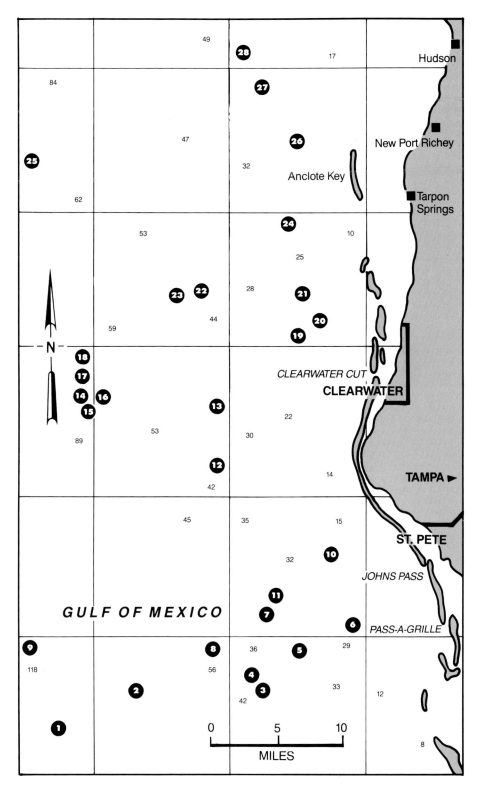

49
28
17
Hudson

84

47
New Port Richey

25
32
Anclote Key
■ Tarpon
Springs

62

53
24
10

25

23 **22**
28
21

44
20

59
19

N

18

17
CLEARWATER CUT

14 **16**
CLEARWATER

15
13

89
53
22

30

12

42
14

45
35
15

32
10

JOHNS PASS

11

GULF OF MEXICO
7

6
PASS-A-GRILLE

9
8
36
5
29

118
56

4
33
12

2
3

42

1

0   5   10

MILES

8

TAMPA ▶

ST. PETE

Area Information

Florida's west coast divers can't boast of tropical coral gardens and crystalline seas in their surrounding waters, but they can boast of some things that will set the spirits of underwater explorers racing: game fish—large, bold and abundant; one of the most extensive and best-maintained artificial reef system available to sport divers; and great wrecks, great spearing, great shelling, and always great adventure.

There are four categories of diving sites in the Gulf: ledges, "drowned" sinkholes, artificial reefs and wrecks. The presence of each breaks the monotony of an otherwise flat, sandy ocean floor. All attract and house complex marine ecosystems with promise of security.

The ledges mark forgotten shorelines, long ago lost to the rising sea. These north-south rock reefs vary in height from one to 12 feet, some extending for miles. The most prominent lines are at depths of 36, 42, 50 and 60 feet. Many are undercut with deep pockets.

"Drowned" sinkholes break the flat submarine plain with limestone shafts that drop vertically to great depths, creating natural fishbowls teeming with sea life.

Artificial reef construction projects were inaugurated to better the marine environment and improve recreation as early as 1961. Close cooperation between county and city governments greatly expanded the scope of the vast undertaking throughout the 70s. Today, their efforts have resulted in a boon for fishermen and divers alike. Sites, carefully chosen with concern for ecology and accessibility, have been built up over the years using solid wastes that stand up to the rigors of the sea. Concrete culverts, building and bridge rubble, steel-hulled ships and barges comprise the bulwark of reef construction materials. Today, however, the marked success of the project is threatened by dredging operations that dump their fill on developed reefs. These reefs, inundated with sludge, cease to exist. The fate of many prolific areas now rests with concerned divers and fishermen.

Wreck diving in the Gulf has always been spectacular, but the additions of the Coast Guard cutter *Blackthorn*, the shrimper *Gunsmoke* and the tug *Sheridan* make the Bay area's list of sunken vessels quite impressive. All are magnificent dives.

Although spring and fall months are considered best for water activities, diving charters make runs during periods of good weather year 'round. Trips are always planned carefully around meteorological reports. Water temperatures vary from the mid-50s in winter to the low 80s in summer months. Water visibility ranges from only a few feet to over 60 feet during periods of calm seas.

Florida spearfishing is at its best in the Gulf. Florida's west coast divers have earned the reputation of being among America's best underwater hunters. National and state spearing competitions are regular occurrences in the region's bountiful waters. Cobia, snapper, bonito, kingfish, mackerel and red and black grouper are their game. Those who gather without spears dive for scallops, oysters, stone crabs and lobster. Here, the shovelnose and slipper lobster are more abundant than the spinies of the east coast and the Keys. Divers should always be ready for adventure when they enter the unpredictably wonderful world of Gulf diving.

NO.1 BIG JACK HOLE

Location: 32 miles from John's Pass on a 225-degree course.

This is one of many "drowned" sinkholes (freshwater sinks that were inundated by saltwater thousands of years ago when the ocean slowly spread its domain across the low, flat Gulf region) that dot the sandy floor of the Gulf.

The entrance (25 feet in diameter) is in 110 feet of water. It plunges straight down to an unknown depth. The limited visibility is totally lost 100 feet down into the shaft. Of course, the depth just at the rim of the hole requires the skills of an experienced open-water diver with advanced training for the depths. Lots of big fish inhabit the area.

NO.2 *GUNSMOKE*   14143.6   44762.4

Location: 24 miles from John's Pass on a 240-degree course.

Great wreck dive! On her final voyage, the 65-foot shrimper lived up to her name. She was scuttled by her crew while the Coast Guard was in hot pursuit. Floating bales of marijuana were all that marked her grave when the cutter arrived. Government divers found only one crew member. He was located below, with a bullet hole through his head. Modern pirates still live by their creed-of-old: "Dead men tell no tales."

The wreck is a beauty. She rests in 80 feet of water, listing slightly to starboard. Shrimp nets remain draped across her rigging.

NO.3 SOUTH JACK WRECK   14137.1   44675.0

Location: 16 miles from John's Pass on a 220-degree course.

The broken remains of an old steel wreck are scattered in 60 feet of water. The intact portion of the wreck, consisting mainly of a large boiler, rests in an upright position on the sand.

NO.4 DOC'S BARGE
Location: 16 miles from John's Pass on a 225-degree course.
The 75-foot-long barge rests in 60 feet of water. She is mostly intact, but split in half.

NO.5 SOUTH JACK LEDGE   14163.5   44678.4
Location: 16 miles out of John's Pass on a course of 220-degrees.
A nice section of ledge, seven to eight feet high, in 50 feet of water. Numerous crevices and undercuts hide large fish.

NO.6 ST. PETERSBURG BEACH REEF   14192.9   44694.1
Location: 5 nautical miles on a 270-degree course from the Pass-A-Grille Channel entrance marker buoy #2.
This area is about 300 feet long with depths of 26 to 28 feet. The first drop was on March 18, 1976; currently there are 151 sections of concrete culverts from the old Cory Avenue bridge on the bottom and sections of the Skyway Bridge. A 200-foot barge was sunk 50 feet east of the center buoy in 1984.

NO.7 TWELVE-FOOT LEDGE
Location: 12 miles from John's Pass on a 240-degree course.

Gulf airplane wreck.   Steve Straatsma

This is the largest ledge in Tampa waters. The 12-foot-high ridge runs for almost one-half mile. Depths are in the 60-foot range.

NO.8 TEN FATHOM WRECK / TRAMP STEAMER   14162.3   44755.8
Location: 16 miles from Pass-a-Grille on a 260-degree course.

The broken remains of a 150-foot tramp steamer lie in 60 feet of water. There is good spearing, with plenty of hogfish and barracuda. Big lobster are found in the summer months.

NO.9 *MEXICAN PRIDE*   14089.6   44898.6
Location: 37 miles west of Pass-a-Grille.

This is a large wreck resting in an upright position in 120 feet of water. It is 80 feet to her top deck. There is good spearing with plenty of red snapper, grouper, large jewfish, cobia, jacks and barracuda.

NO.10 TREASURE ISLAND REEF   14200.8   44738.7
Location: 4.8 nautical miles on a 248-degree course from the Johns Pass entrance marker; or 7.2 nautical miles on a course of 304 degrees from the Pass-A -Grille entrance buoy #2.

The reef is marked on each end by buoys. Depths range from 29 to 33 feet. The first drop was on January 23, 1976; currently in place are 40,000 car tires, 1,032 truck tires, and 561 sections of concrete culvert. Many black grouper and mangrove snapper. Pyramid-shaped structures with 12-25-foot profiles, and diameters of 50-65 feet are located near the south, center and north buoys.

NO.11 MECCO'S BARGE/*BETTY ROSE*   14184.3   44769.2

Location: 10 miles from John's Pass on a 245-degree course.

A 75-foot barge, completely intact, rests in an upright position in 45 feet of water.

NO.12 INDIAN SHORES REEF   14200.0   44859.7

Location: 13.6 nautical miles on a course of 292-degree from Johns Pass entrance marker; or 11.9 nautical miles on a 235-degree course from the entrance bell marker in Clearwater Pass.

Depths are 45 feet in this artificial reef site. 125 pillboxes were the first placed here in 1962. A 235-foot Navy Landing Ship, filled with cable, was sunk in January 1976. It is located 100 feet east of the south buoy. A second LSM is 100 feet west of the center buoy. A 240-foot salt hopper barge is lying upside-down 200 feet southwest of the north buoy.

NO.13 PINELLAS #1 / RUBE ALLYN ARTIFICIAL REEF   14212.3   44886.4

Location: 9.8 nautical miles on a 256-degree course from the Clearwater Pass entrance bell marker #1.

A large barge was sunk on August 11, 1976. It is located 100 feet east of the center buoy. A pyramid-shaped structure constructed of plastic covered with fiberglass, called a Japanese fish attractor, is located 200 feet due north of the north buoy. It must be working; a world's record cobia of 88 pounds was caught here in 1982! Visibility averages 25 feet in summer. Bottom depths are 50 feet.

NO.14 *BLACKTHORN*   14181.7   44942.6

Location: 30 miles from John's Pass on a 292-degree course. inside the Pinellas # 2 Artificial Reef Site. Marked by an orange buoy.

The wreck of the 110-foot U.S. Coast Guard cutter *Blackthorn* is one of the Gulf's most popular dives. The *Blackthorn's* fate was international news in late '81, when the cutter was broadsided by a freighter in one of the channels of Tampa Bay. Twenty-two of her crew members lost their lives in the collision. Her hull, damaged far beyond repair, was towed to thePinellas #2Artificial Site and scuttled.

The huge mass of steel quickly attracted sea life as well as hundreds of divers each year. The rather shallow depths of the large wreck (30 feet to her upper decks; 80 feet to the ocean floor) afford an opportunity for less experienced divers to visit a wreck. Penetration of her inner chambers is not recommended because of the

Arrow Crab   Ned DeLoach

inherent dangers of all intact wrecks (lack of light; sharp, broken fittings; mazes of passages that easily disorient the diver; and silt). The broken remains of a 250-foot barge, tires and other debris lie nearby **14181.6 44943.3.**

NO. 15 TUG *SHERIDAN*   14181.9 44941.8
Location: 100 yards southeast from the marker on the Blackthorn in the Pinellas #2 Artificial Reef Site.

A hot new dive! The *Sheridan* is a 180-foot ocean going tug sunk resting in 75 feet of water. She sits upright with her prop intact. The tug is surrounded by concrete culverts and tires.

NO.16 TABLE TOP   14198.2 44913.0
Location: 1/2 mile east of the Blackthorn.

The Table Top is a plateau over 200 feet in diameter that rises seven to eight feet from the sand. Large undercuts filled with sea life are interesting to explore.

NO.17 THE CAVES
Location: 1/4 mile northwest of the Blackthorn.

The caves are actually deep undercuts in an eight-foot-high ledge. This is a good area for underwater hunting. Depths go to about 80 feet.

NO.18 CLEARWATER WRECK   14203.8 44944.9
Location: 23 miles from entrance bell marker #1 in Clearwater Pass on a 267-degree course.

The wreckage of a large steamer lies in 60 feet of water. Her hull is split in half and rises only 20 feet off the flat sea floor.

NO.19 'G' MARKER
Location: 4-1/2 miles from bell marker #1 in Clearwater Pass on a 295-degree course.

This is a nice close-in area, best dived on calm, clearwater days. The broken rock ledges run for nearly half a mile, rising about five feet off the bottom. This is a good spot to see large marine life close to shore. Many jewfish, nurse sharks and turtles have been spotted. The depth averages 25 feet.

NO.20 CLEARWATER REEF   14233.3   44851.1
Location: 3.7 nautical miles on a 316-degree course from the Clearwater Pass entrance bell marker 31; or 3.3 nautical miles on a 265-degree course from the Dunedin Pass entrance; or 4.2 nautical miles on a 237-degree course from the Hurricane Pass entrance marker #2.

A large area marked by four buoys. Clearwater Reef is one of the largest and most popular artificial reef sites in the area. Depths range from 27 to 29 feet. The reef was started on June 2, 1965 with an initial drop of 75 specially constructed concrete pillboxes; additional drops provided at least 45,000 tires. Several steel barges are located in the center of the area. Although this is the oldest reef, the marine population was killed during a red tide outbreak in August 1974. The reef cycle began to renew after the kill. A large variety of tropicals make their home among the rubble. Game fish such as snapper and hogfish are common, as well as some lobster. Spearfishing is popular.

NO.21 DUNEDIN REEF   14247.9   44887.3
Location: 6.3 miles from the entrance bell marker #1 in Clearwater Pass on a 326-degree course; or 4.4 miles on a 270 degree course from the Hurricane Pass entrance marker #2.

This artificial reef area is marked by buoys on its northern and southern ends. The northern buoy is near a natural rock ledge. Concrete culverts are scattered to the south for 300 feet. There are plenty of game fish, and snook are often seen in the summer months. Depths range from 25 to 30 feet.

NO.22 BOMBER
Location: 14 miles from Tarpon Springs on a 295-degree course.

The broken wreckage of what was probably a WWII transport lies in 50 feet of water. Only the fuselage remains intact.

NO.23 MASTHEAD LEDGE
Location: 16 miles from bell marker #1 in Clearwater Pass on a 285-degree course.

A tall, long ledge section that rises eight feet from the bottom and runs for over a mile and a half, this is a popular dive because of the formation's size and the extensive marine growth of the rock outcropping. There are many deep crevices and undercuts offering plenty of hiding places for marine life. Shelling is good along the ledge. Helmet shells and large horse conch are commonly found.

NO.24 TARPON SPRINGS REEF   14259.3   44935.3
Location: 12 nautical miles from the entrance bell marker #1 in Clearwater Pass on a 338-degree course; or 7.6 nautical miles on a 312-degree course from the Hurricane Pass entrance marker #2; or 3.7 nautical miles on a 270-degree course from the south entrance marker 31 in the Anclote Anchorage.

The south buoy marks the beginning of the artificial reef. The north buoy is over natural rock ledges. The artificial reef material, including concrete culverts, starts north of the southern buoy. Depths range from 26 to 28 feet.

NO.25 TUGBOAT AND BARGE
Location: 30 miles west of Tarpon Springs.

A large 105-foot tug and the 80-foot barge she was towing went down in 85 feet of water during high seas. The tug, mostly intact, rests upside-down on her superstructure. A jeep lies nearby. The barge came to rest less than a mile to the north.

NO.26 HELLCAT
Location: Between 3 and 4 miles from the north end of Anclote Key on a 270-degree course.

A WWII Hellcat rests in 25 feet of water. She remains pretty much intact with wings and fuselage in place. There are no ledges in the area.

NO.27 PASCO COUNTY ARTIFICIAL REEF 'PS'   14275.4   44997.5
Location: 11 miles west of Gulf Harbor in New Port Richey on a 270-degree course.

Four 200-foot barges were sunk in 25 feet of water. One barge is at buoy 'P'; another, 1,500 feet directly north at buoy 'S'. Cement culverts are scattered between the middle barges. There is a lot of fish activity around the wrecks. Jewfish, cobia, sheepshead and snapper frequent the area. Flounder are common in flat sand areas surrounding the reef. There is good spearfishing here.

NO.28 PASCO COUNTY ARTIFICIAL REEF 'CO'   14274.9   45048.6
Location: 15 miles from Gulf Harbor in New Port Richey on a 284-degree course.

The broken remains of a barge lie near buoy 'C' in 30 feet of water. Several two-foot-high rock ledges run near buoy 'O'. The area is alive with fish life. Lobsters are often pulled from the ledges. Sea whips, sponges and fire coral are common.

12
BRADENTON
SARASOTA - VENICE

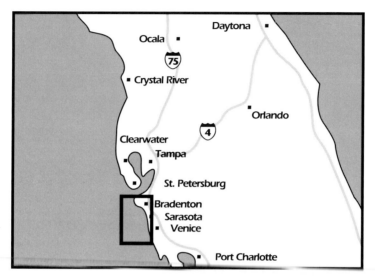

Offshore Diving Locations

1. Sugar (Molasses) Barge
2. Third Pier
3. Barracuda Hole
4. Gully's
5. Barge and Hopper
6. Venice Ledges
7. Venice Public Beach
8. The Rocks
NOT ON MAP:
   Bay Ronto
   Peace River

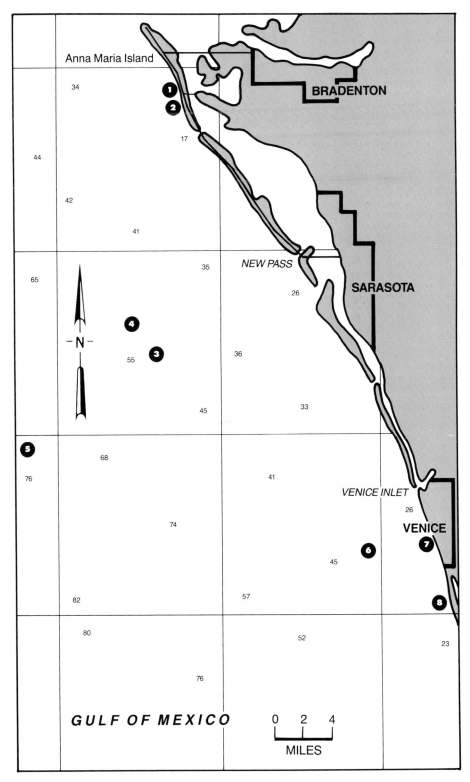

Anna Maria Island

BRADENTON

34

44

42

41

17

35

NEW PASS

65

26

SARASOTA

4

3

55

36

45

33

5

68

76

41

VENICE INLET

74

26

VENICE

6

7

45

8

82

57

80

52

23

76

GULF OF MEXICO

0 2 4

MILES

Area Information

Although huge jewfish are commonly sighted at many dive sites in the Gulf, Florida's southwestern coastal zone is called "Jewfish Country" because of their proliferation on the local ledges and wrecks. Few dives are made without spotting one or more. Often, over 20 of the gigantic fish are sighted on a single descent. The view of such a massive creature casually gliding through its sovereign domain is long remembered.

Spearfishing is generally the activity of the day. Snapper and grouper are the prime game. Another popular pastime is searching for large, prehistoric sharks' teeth just off the Venice public beach or on the muddy bottom of the Peace River. The abundance and quality of the teeth rival those found in the Cooper River in South Carolina.

NO.1 SUGAR (MOLASSES) BARGE (Beach Dive)

Location: About 100 yards straight out from Trader Jack's Restaurant on Bradenton Beach. A metal post on the barge's bow protrudes from the water, marking the exact location.

This old, 75-foot barge has long been a popular dive because of its close proximity to shore and its shallow depths (15 to 20 feet). If you catch a day with good visibility, you'll be delighted with the variety of sea creatures that make the wreck their home. Colorful soft corals and sponges are common.

NO.2 THIRD PIER (Beach Dive)

Location: About 200 yards off Bradenton Beach. The entry site is adjacent to the intersection of Gulf Drive and 33rd Street. While swimming out, bear slightly south.

This is the best close-in ledge area on the Gulf coast. Depths vary from 20 to 30 feet. The broken reef line runs all along the coast to Trader Jack's Restaurant. Best diving conditions exist when the wind is out of the east. Westerly winds can reduce water visibility to zero.

NO.3 BARRACUDA HOLE

Location: 9.5 miles from the New Pass sea buoy on a 240 degree course.

This area is known for its many rocky outcroppings and ledges up to six

feet high. The bottom is in 50 feet of water. A large variety of fish are present, including snapper, amberjack, grouper and jewfish. Sponges and soft coral adorn the ledges.

NO.4 GULLY'S

Location: 11 miles from the New Pass buoy on a 250-degree course. The area is just northwest of Barracuda Hole.

This is a ledge area pocketed with undercuts where depths range from 54 to 56 feet. The bottom is generally rocky, with a few hard coral formations. Spiny lobster are occasionally pulled from the rocks. Shovelnose lobster are more common and can be taken year-round. A lot of tropical and game fish inhabit the ledges. Grouper, mangrove snapper, and hogfish are the usual catch-of-the-day for spearfishermen.

NO.5 BARGE AND HOPPER   14128.1   44495.4

Location: 19 miles southwest of the New Pass sea buoy on a 238-degree course.

This is a 100-foot barge resting upside-down in 65 feet of water. It has large breaks in the hull which lead inside. It is dangerous, however, to venture inside due to the weak structure of the collapsed sections. Large mangrove snapper can be found in and around the northern end of the barge. Jewfish, barracuda and amberjack are common.

The hopper is half a mile south of the barge. It is draped with nets and lines, and it would be wise to stay clear due to the possibility of entanglement. This is

a good spot to grab a lobster. It has been reported that the hull and one wing of a large airplane are located near the hopper.

NO.6 VENICE LEDGES

Location: A large area extending from 1 to 5 miles offshore. Located out from the Venice jetties on a 210-degree course, or directly west of the Venice fishing pier.

Depths range from 20 feet in the area close to shore to 50 feet on the outer fringes. The rocky bottom terrain and small ledges are popular spearfishing grounds. Visibility is moderate, but extends to 30 feet on good days. Large black grouper and jewfish are spotted on every dive. Large turtles frequent the area most of the year.

NO.7 VENICE PUBLIC BEACH (Beach Dive)

Location: 30 to 40 yards off the Venice Public Beach, in 18 feet of water.

The area off this section of beach contains one of the largest deposits of prehistoric sharks' teeth ever found. Specimens of up to six inches across have been discovered, while smaller ones are quite common. They can be spotted on the bottom and uncovered by gently fanning the sand with your hand. Visibility of 3 to 4 feet can be expected on calm days. The best spots seem to be at the exact depth of 18 feet. Conditions are best during periods of easterly winds.

NO.8 THE ROCKS (Beach Dive)

Location: From Route 776 in south Venice, take the Manasota Beach Road to the public beach area. It is a 1-1/2 mile walk south to the site, which is located halfway between Manasota and Middle Road. Unfortunately, there are no closer access roads because the adjacent property is all privately owned.

A large, flat, rocky area that runs for about one mile and extends 75 yards offshore gives this site its name. Depths range from one to 18 feet. Visibility averages 2 to 5 feet, but reaches over 20 feet a couple of times a month. Large, flat, moss-covered rocks are closer to shore. The bigger rocks are in deeper water. Snook, snapper and grouper are frequently sighted.

BAY RONTO (Not on Map)   14140.9   44325.1

Location: 30 miles from the Venice jetties on a 224-degree heading.

The *Bay Ronto* is a super dive, but should be limited to highly experienced open-water divers because of its depth of 110 feet. In 1919, the 400-foot German freighter went down in stormy seas when her cargo of grain shifted. She rests upside-down on a sandy floor. Her superstructure is not with the wreck. There are many areas through which the broken hull of this gigantic wreck can be entered. The site is rated excellent for the experienced spearfisherman and for underwater photographers. Recently, her huge bronze propeller was removed, lessening the photogenic aspect of the hull. Visibility is usually excellent, due to the water depth and distance from shore. Large amberjack and barracuda linger above the wreckage, while several large jewfish have taken up permanent residence inside her hull.

PEACE RIVER (Not on Map)

Location: Travel west for 41 miles from Sarasota to Arcadia on S-72 (Bee Ridge Road). Go four miles on US 17 to the small town of Nocatee; then go east on C-760 to the river.

The Peace River has long been South Florida's most popular river recreational area. Canoeing and camping are excellent. This shallow, darkwater river is dived for only one reason—large prehistoric sharks' teeth. Use only small boats to get about because the river is shallow, usually 3 to 4 feet deep. Canoes can be rented at the Canoe Outpost in Arcadia.

One of the best spots for hunting is found at the river's double bend, one-quarter to a half mile south of the Nocatee Bridge. Diving conditions are best during the winter, when the water level is low and the gators are not so active. Water visibility is nonexistent. The hunter must carefully feel around in soft mud. Be sure to wear gloves. Thousands of large, well-preserved sharks' teeth and arrowheads have been found using this method.

13
PORT CHARLOTTE
PUNTA GORDA - FT. MYERS

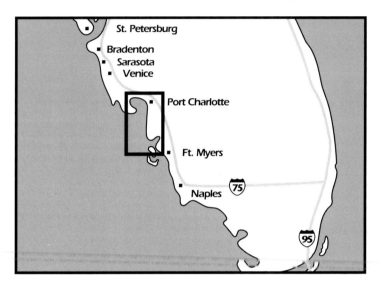

Offshore Diving Locations

1. Boca Grande Causeway
   Bridge & Railroad Trestle
2. Seventeenth Street Reefs
3. Boca Grande Jetties
4. Dock Wreck
5. *Cathy II*
6. The Gardens
7. Mud Hole
8. The "W"
9. Pinnacles
NOT ON MAP:
   Mine Sweeper Wreck

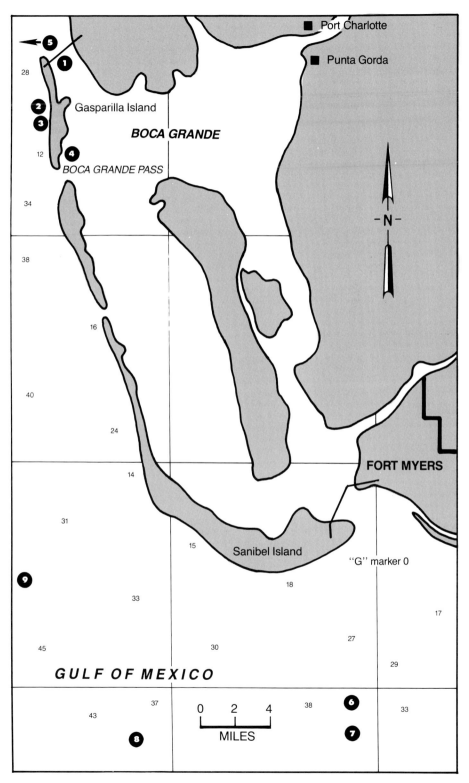

■ Port Charlotte

■ Punta Gorda

Gasparilla Island

BOCA GRANDE

BOCA GRANDE PASS

28

2

3

12

4

34

38

16

40

24

14

31

15

Sanibel Island

9

33

45

30

GULF OF MEXICO

37

43

8

-N-

FORT MYERS

"G" marker 0

18

17

27

29

0 2 4

MILES

38

6

7

33

Area Information

This area has long been famous for the prime shelling grounds found along the beaches of the coastal islands of Sanibel and Captiva. Unfortunately, the visibility in the surrounding waters is very poor due to the muddy runoff from the Peace River. Gasparilla Island, located just north of Charlotte Harbor, is an exception. Here, visibility varies from three to 15 feet. Spearfishing and gathering stone crab claws are popular activities. It is extremely dangerous to dive in the cuts between any of the islands due to strong tidal currents.

NO.1 BOCA GRANDE CAUSEWAY BRIDGE AND RAILROAD TRESTLE (Beach Dive)
Location: Travel 4 miles on US 41 from Port Charlotte. Turn west on SR 771. Go about 15 miles to the toll bridge over the Intracoastal Waterway. The bridge will take you to the small town of Boca Grande on Gasparilla Island. Tolls are $3.00 for trucks or vans; $1.50 for cars.

Good spearing and stone crabbing can be found around the support structures of the toll bridge and adjacent railroad trestle. Depths range from 20 to 25 feet. It is best to dive the area one-and-a-half hours before high tide. You can expect three- to five-foot visibility. There are a lot of grouper and a few snapper in the area. Tide charts can be picked up at local dive shops

NO.2 SEVENTEENTH STREET REEFS (Beach Dive)
Location: Off the west end of 17th Street in Boca Grande. Bear slightly to the right as you swim out from shore.

Spadefish on a Gulf wreck.   Steve Straatsma

The first group of rock ledges is in five to eight feet of water. These ledges are about a foot high. One hundred yards out, in 15 to 20 feet of water, the ledges are three- to four-feet high.

NO.3 BOCA GRANDE JETTIES (Beach Dive)
Location: On the west (ocean) side of the island, about 1 mile from the south end.

Three rock jetties extend out into the Gulf. Depths range from 10 to 12 feet. There are few fish but plenty of stone crabs.

NO.4 DOCK WRECK (Beach Dive)
Location: In the Intracoastal Waterway, just north of the large phosphate docks. The broken remains of an old shrimper lie scattered over the sand bottom.

NO.5 *CATHY II*
Location: 5 miles west of the north end of Boca Grande.

The twisted hull of a 65-foot shrimper rests in 42 feet of water. The gas tanks and engine are still intact. Visibility varies from 20 to 40 feet.

NO.6 THE GARDENS
Location: About 7 miles from FL G "1" buoy, off Ft. Myers Beach on a 210-degree course.

The Gardens is a large, scenic area with a lot of soft coral and fish life, including plenty of tropicals. Visibility ranges from 30 to 40 feet; depth averages 25 feet.

NO.7 MUD HOLE
Location: About 9 miles from FL G "1" buoy, off Ft. Myers Beach on a heading of 197 degrees.

Mud Hole is a poor choice of name for this live freshwater spring. The water issues from a fissure two-feet wide. On a calm day, the spring can be spotted by change of water color and boil on the surface. The depth is 40 feet. Visibility averages 20 feet. Very large jewfish and sharks inhabit the area.

NO.8 THE "W"
Location: About 16 miles from FL G "1" buoy off, Ft. Myers Beach on a 240-degree course.

"W" stands for the shape of the rock reef that is the favorite hunting ground of local spearfishermen. Depths vary over the large bottom area from 45 to 48 feet. Good sponge and gorgonian growth is found on the rocks. Visibility averages 25 feet.

NO.9 PINNACLES
Location: 22 miles due west of FL G "1" buoy, off Ft. Myers Beach.

The Pinnacles consists of five or six very interesting rock formations which resemble steeples, rising 15 feet from the ocean floor. Depth averages 55 feet around the spires. Visibility averages 30 feet.

MINE SWEEPER WRECK (Not on Map)
Location: 25 miles from FL G "1" buoy, off Ft. Myers Beach on a 200-degree course.

This 60-foot wreck rests in 70 feet of water. Snapper and grouper hole up in the

14

NAPLES

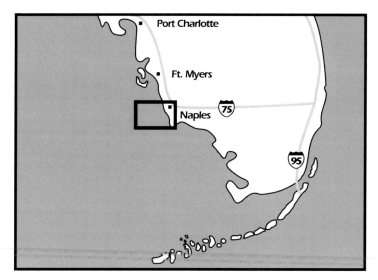

Offshore Diving Locations

1. Black Hole
2. House Boat Wreck
3. Mine Sweeper
4. Naples Ledges
5. Crater

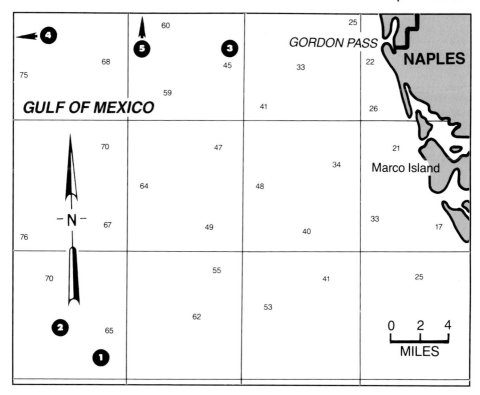

Area Information

If you have a penchant for adventure when you dive, head for Naples. Here the Gulf sea covers an abundance of exciting bottom yet to be explored by man. Unruly winter seas usually limit diving to the spring and summer months. Runs to prime diving grounds are long and visibility is limited, but, when you do get down, be ready for a thrill. Fish are large and abundant, turtles plentiful and shelling superb. Both ledge reefs and wrecks support a profusion of large colorful sponges and gorgonians. The large jewfish colonies that inhabit area wrecks and ledges are impressive.

NO.1 BLACK HOLE   14028.6   43864.2
Location: 24 nautical miles southwest of Gordon Pass.

The circular entrance to this large 'drowned' sink is 100 feet in diameter. The rim is in 65 feet of water and plunges to a 215-foot depth. The limestone shaft takes on an hourglass shape at 115 feet. Fish life is plentiful with large jewfish and grouper common. Sea turtles frequent the area. Several sponge-covered ledges can be found nearby.

Ned DeLoach

NO.2 HOUSE BOAT WRECK

Location: 27 nautical miles southwest of Gordon Pass, or about 4 miles north of the Black Hole.

This is a great spot to observe large jewfish. Many make their homes in and around the 50-foot wreck. The boat remains are intact and rest upside-down in 70 feet of water.

NO.3 MINE SWEEPER

Location: 11 nautical miles west of Naples.

The vessel is mostly broken apart. Water visibility is generally less here than at other area sites, but the large concentration of game fish makes it a prized dive for local spearfishermen. Depths are to 45 feet.

NO.4 NAPLES LEDGES

Location: 30-40 nautical miles west of Naples.

Several ledge lines run parallel to one another for miles. Depths range from 70 to 80 feet. Many virgin areas can still be found along the line, and several varieties of snapper and grouper inhabit the seven- to eight-foot-high ledges.

NO.5 CRATER

Location: 22 nautical miles northwest of Naples.

The Crater resembles a large bowl on the ocean floor. The 15-foot depression is brimming with game fish, making it a spearfishermen's delight. 'Cuda and snapper abound.

15

JACKSONVILLE

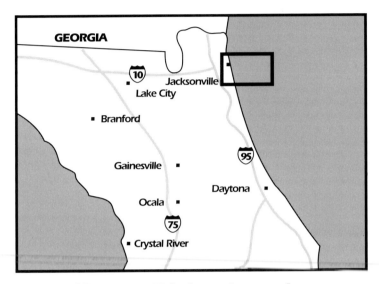

Offshore Diving Locations

1. Montgomery's Reef
2. Rabbit's Lair
3. Nine Mile Reef
4. Paul Main's Reef
5. Amberjack Hole
6. East Fourteen
   (Gator Bowl)
7. Middle Ground
8. Clayton's Holler
9. *Anna*

10. Coppedge Tug
11. Southeast 16-17
12. Hospital Ground
13. Blackmar's Reef
14. *Casa Blanca*
15. *Hudgins*
NOT ON MAP:
   Dry Dock
   Lost Phantom

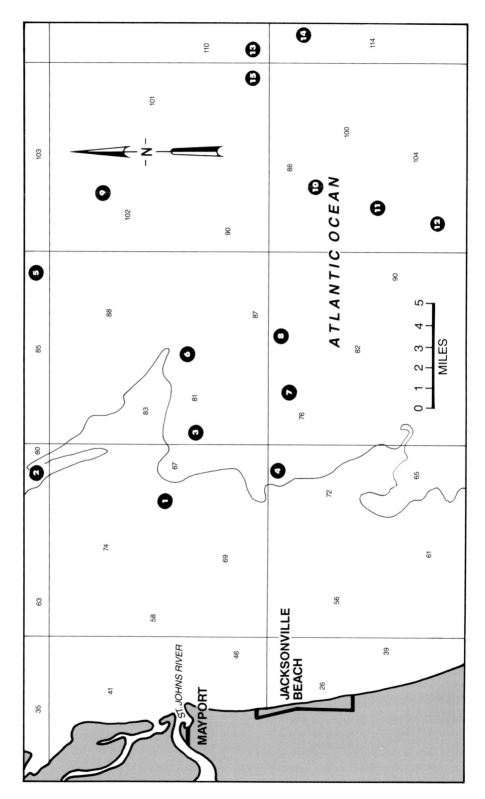

Area Information

Offshore Jacksonville has long been the center of diving along Florida's northeast coast. The extensive system of limestone reef ledges, coupled with 25 years of intense artificial reef building, has produced an ocean floor brimming with dramatic seascapes. What makes these waters so enticing, however, is the abundance of marine life. Each ledge and artificial reef swirls with fish. For two decades spearfishermen were the only divers who witnessed the phenomenon. Recently, dive boat charter operations opened the area to sport divers. From spring through the summer and into early fall, they make regular runs to the wrecks and reefs. Divers return daily with wild and wonderful stories about their exciting adventures in the "sea of fish."

The best conditions are found during the summer months when calm, warm, clear seas prevail. Visibility varies between 20 and 120 feet; typically it improves as you get further offshore. Depths are from 60 to just over 100 feet.

NO.1 MONTGOMERY'S REEF   45232.7   61958.9
Location: On a course of 73-degrees, 8.4 nautical miles from the Mayport jetties.

Over a dozen shipwrecks, surrounded by several small reefs, support many grouper to 20 pounds, along with snapper, amberjack, cobia and sheepshead. The small reefs are excellent for shelling and tropical fish collecting. The 63-foot tug *Reliance* (the Loran number above is for this wreck), "Culvert Reef," and "Thunderdom" are popular with spearfishermen. Depths average 65 feet, with summertime visibilities averaging 50 feet.

NO.2 RABBIT'S LAIR   45248.6   61914.1
Location: On a course of 57-degrees, 13.1 nautical miles from the jetties.

Several long ledges, up to ten feet in height, provide an excellent area for spearfishing and lobstering. A nearby mine sweeper, wooden tug boat and sunken barge are homes to many varieties of game fish. Depths average 70 feet with a typical visibility of 30 feet. The Loran number above is for a high section of ledge, the outstanding part of this site.

NO.3 NINE MILE REEF   45192.0   61944.8

Location: On a course of 96-degrees, 10.4 nautical miles from the jetties.

This area is popular for its wrecks. Four of the most often dived sites are the "Open-Sided Barge," "Vic's Barge," the "Asphalt Barge"(The Loran number above is for this barge which is roughly in the center of the site.), and an old 52-foot steamer tug. These wrecks hide grouper, averaging 15 pounds, as well as large flounder and sheepshead. The steamer tug is excellent for artifact hunting. Nearby reefs offer good photographic opportunities. Depths average 75 feet, with visibilities around 30 feet.

NO.4 PAUL MAIN'S REEF   45174.0   61964.9

Location: On a heading of 115-degrees, 10.6 nautical miles from the jetties.

In addition to several small reefs, this area has two photogenic tugboats sitting upright on the bottom. Other wrecks include the "Porgy Boat" and the "Banana Boat," which are next to one another. Ten to 15-pound grouper are common. Summertime visibility averages 30 to 40 feet; depths average 75 feet.

NO.5 AMBERJACK HOLE   45209.9   61849.9

Location: On a course of 64-degrees, 18.8 nautical miles from the Mayport jetties.

Amberjack Hole is Jacksonville's most beautiful natural reef. The high 10-foot ledge is completely encrusted with soft corals and sponges. Its many crevices and caves hide crustaceans and tropical fish. Schools of amberjack are commonly seen charging down the reef line, cutting through the haze of baitfish. Truly a photographer's delight. There are several other small reefs in the area. The depths vary from 75 to 85 feet; visibility averages 40 feet.

NO.6 EAST FOURTEEN (GATOR BOWL)   45149.9   61887.5

Location: On a course of 90-degrees, 15 miles from the jetties.

This area includes a very popular dive called the "Gator Bowl Pressboxes." (The Loran number above is for the center of the pressbox.) Over a dozen reefs attract huge schools of amberjack and spadefish. "Red Rock Reef," a beautiful soft coral cliff, is a popular night dive **45145.4 61873.4**. The "Y-Ledge" has several caves to hold game fish **45152.7 61891.4**. Visibilities average 40 feet, with depths to 85 feet.

NO.7 MIDDLE GROUND   45130.6   61937.3

Location: On a course of 117-degrees, 13.5 nautical miles from the jetties.

The Middle Ground is made up of a series of small reefs and two popular artificial reefs. A steel tug sits at a 45-degree list in 85 feet of water. (The Loran number above is for this tug.) It is encrusted with a thick layer of soft corals and sponges. Wreckage from the wheelhouse lies alongside providing good cover for grouper and sheepshead. The "Japanese Reef" was placed as part of an artificial reef building project funded by the Japanese government. This experimental reef is a pyramid of fiberglass tubes made of interlaced hoops. It has been very effective in attracting pelagic fish **45141.4 61946.1**.

NO.8 CLAYTON'S HOLLER   45132.0   61916.1

Location: On a course of 111-degrees, 16.7 miles from the Mayport jetties.

Clayton's Holler is the most popular Jacksonville diving area. There are three distinct groups of reefs, each with a dozen ledges. The ledges in the north group run parallel (The Loran number above is for this section.); one has a large swim-through in its center. The middle area is a mile-long reef **45117.3 61914.6.** Two tug boats and a barge sit just off the reef. The southern group has the tallest offshore reef, rising 15 feet off the bottom **45114.2   61923.1.** Depths range from 85 to 95 feet; visibility averages 40 feet.

NO.9 *ANNA*   45128.8   61808.8

Location: On a course of 78-degrees, 22.2 nautical miles from the jetties.

This 180-foot freighter sunk in 1986 is now a popular dive site. The wreck is lying on its side. Its bow and stern are still intact, but the cargo hold has been broken down into a stack of plates, perfect for hiding game fish. A picturesque tug, a large barge and a small reef area, known as the "Lobster Hotel," are nearby **45151.4 61800.0.**

NO. 10 COPPEDGE TUG   45065.7   61866.0

Location: On a course of 107-degrees, 22.6 nautical miles from the jetties.

Since it was sunk in 1988, this 75-foot tug has attracted an unusually large amount of sea life. Because there are no natural ledges in the area the steel structure has drawn fish like a magnet. When you first approach the wreck, the pilot house is virtually obscured by the clouds of baitfish. Even though the site is over 20 miles offshore, the depth is only 75 feet. This makes the wreck an ideal second dive after visiting deeper ledges in the area. The tug is intact and sits upright with the rudder and prop still in place. Just east of her bow is a series of culverts and pillboxes. These are also covered with fish.

NO.11 SOUTHEAST 16-17   45054.2   61876.2

Location: On a course of 115-degrees, 22 nautical miles from the jetties.

This large area of natural reefs includes the popular "19 Fathom Hole." The reefs vary from 10-foot-high ledges to an area of broken bottom covered with baitfish. The "Octopus Garden" is a limestone reef riddled with holes occupied by large octopuses. Depths range from 95 feet to 105 feet, while visibility can reach 90 feet.

NO.12 HOSPITAL GROUND   45050.5   61900.5

Location: On a course of 122-degrees, 22.1 nautical miles from the jetties.

Another large area of natural reefs similar to Southeast 16-17. There are several tall ledges with deep caves that hold grouper and snapper. An excellent area for lobster. Depths average 95 feet; summertime visibility averages 50 feet.

NO.13 BLACKMAR'S REEF   45047.5   61783.7

Location: On a 95-degree course, 25.2 miles from the jetties.

Five large wrecks and several small reefs make up the Blackmar. The "Warwick" ferry boat (the Loran number above is for this wreck), the "Ocean Going Tug," **45050.7   61788.7** and the "Super Barge" **45050.3   61785.3** are all good places to spearfish. All the wrecks house tremendous schools of spadefish and amberjack. Grouper and snapper are common; you may even see jewfish. A Banshee jet fighter and a WWII Corsair are also in the area. Both are sitting intact on the bottom and make thrilling photo dives. The small reefs and soft coral bottom hide six- to eight-pound lobster. Depths range from 95 to 110 feet. It's not uncommon to find 100-foot visibility.

NO.14 *CASA BLANCA*   45012.7   61795.2

Location: On a course of 105-degrees, 29.4 nautical miles from the jetties.

This LST landing craft is an impressive wreck. It sits upright on the bottom in 112 feet of water with the wheelhouse and upper deck 85 feet below the surface. Schools of amberjack and spadefish swim in spirals from the wreck to the surface. Large

Ned DeLoach

grouper and snapper are consistently found in the holes and under broken plates. Visibility this far offshore often reaches 100 feet.

NO.15 *HUDGINS*   45077.3   61813.9
Location: On a course of 97-degrees, 24.7 nautical miles from the jetties.

The *Hudgins* is an intact 150-foot freighter sitting upright on the bottom in 105 feet of water. She was put down in 1987. Schools of spadefish and amberjack reach to the surface. Baitfish and snapper cover the wreck like a moving cloud. A great spot for underwater photography! Several small reefs and a barge are in the area. Summertime visibility averages 50 feet.

DRY DOCK (Not on map)   44839.4   61710.9
Location: On a course of 111-degrees, 46 nautical miles from the jetties.

A huge dry dock from the Jacksonville Shipyards was sunk in 1989. This is the largest artificial reef in Florida. The structure is 600 feet long and over 400 feet wide. Soon after settling to the bottom the side walls collapsed out to the sides. Four structures 50 feet long and 20 feet wide tower up 50 feet at the corners. These support a series of catwalks and ladders. The dock rests in 127 feet of water. It is a 110-foot dive to the floor and 75 feet to the top of the corners. The site almost immediately became a fishbowl. Cobia, blue runners and other jacks are everywhere. Grouper and jewfish hide along the edges. Because of its distance from shore the visibility is usually 100-foot plus. This dive is well worth the long boat ride.

THE LOST PHANTOM (Not on map)
Location: On a course of 116 degrees, 41 miles from the Mayport jetties.

An F4-A Phantom jet fighter sits upright in the sand with cable tangled around its landing hook and the canopy blown off. This is one of the most thrilling airplane wreck dives off Jacksonville. A huge school of amberjack inhabit the area and jewfish commonly make their home under the tail section. Visibility is usually in excess of 100 feet. The depth is 102 feet.

16
DAYTONA BEACH

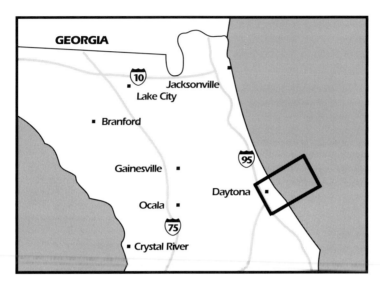

Offshore Diving Locations

1. Liberty Ship *Mindanao*
2. Twin Airplane Wreck
3. Port Orange Bridge Rubble
4. Nine Mile Hill
5. Twelve Mile Wreck
6. North East Grounds
7. Cracker Ridge
8. Party Grounds

9. Culvert Reef
10. East Eleven
11. Turtle Mound
12. Freighter Wreck
13. North Jetty
NOT ON MAP:
    Rainbow Reef
    Sinkhole off Flagler

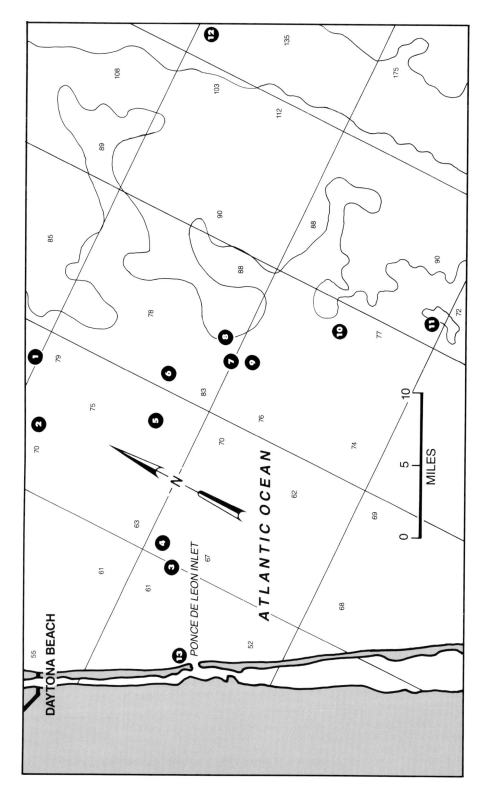

Area Information

The waters off Daytona Beach offer a variety of challenging diving experiences, from spearfishing and "bug snatchin" to wreck diving on WWII planes and Liberty ships. They even conceal a 19th century Spanish-American War gunrunner. The summer months are best for underwater adventure when the sea tends to calm, and the visibility fluctuates between 25 and 75 feet, and its temperature hovers at 80 degrees. Like the other areas of northeast Florida, there is not any good beach diving in Daytona. All the wrecks, artificial reef sites and limestone ledges are from eight to 20 miles offshore.

NO. 1 LIBERTY SHIP *MINDANAO*   44453.8   61982.3
Location: 10.6 nautical miles on a 48-degree course from the sea buoy at the Ponce DeLeon Inlet.

In 1980, the Halifax Sport Fishing Club sank the 446-foot Liberty ship *Mindanao* in 80 feet of water. The large wreck is intact but has a break across its 62-foot hull from its impact with the ocean floor. The bow is pointing north. Before sinking, large holes were cut in all major compartments to allow access for fishlife and divers. Entries can be made into the brig, engine room and crew's quarters. The vessel is three deck high. The flat top deck is 50 feet below the surface.

NO.2 TWIN AIRPLANE WRECK
Location: 16 nautical miles on a 25-degree course from the sea buoy at Ponce DeLeon Inlet. Local guides should be contacted for exact location.

This is an area where two TB-M torpedo bombers crashed during WWII. The wrecks are located approximately 200 yards apart, in 71 feet of water. Excellent spearfishing is to be had here, with many snapper and large sheepshead.

NO.3 PORT ORANGE BRIDGE RUBBLE   44437.8   62020.0
Location: 6.7 nautical miles on a 64-degree course from the sea buoy at the Ponce DeLeon Inlet.

In 1989, the remains of the Port Orange and the New Smyrna Bridges were put down in 75 feet of water. The concrete and metal guard railings rise 15 to 18 feet off the bottom.

NO.4 NINE MILE HILL   44473.4   62035.3
Location: 6.5 nautical miles on a 43-degree course from the sea buoy at Ponce DeLeon Inlet.

This is a large artificial reef composed of old cars, tires and concrete pilings. There is also some natural reef in the area with four-foot ledges. However, it is difficult to locate. The artificial part of the reef is clearly marked with two large orange balls, one at each end.

NO.5 TWELVE MILE WRECK   44281.2   61736.6
Location: 10.8 nautical miles on a 52-degree course from the sea buoy at Ponce DeLeon Inlet. Marked by a black flag during the summer.

This wreck sank during the 1800s, and is reported to be an old Spanish-American War gunrunner, deduced from the large amounts of Gatling ammunition found around it. Large jewfish inhabit the area along with giant amberjacks and vast numbers of spadefish. This is an unusually beautiful dive during the summer when the water is clear. Local guides must be contacted for assistance in locating this small wreck.

NO.6 NORTH EAST GROUNDS   44432.8   61932.0
Location: 13 nautical miles on a 45-degree course from the sea buoy at Ponce DeLeon Inlet. The area is marked by a black flag, with NE on the float.

The North East Grounds is a good area for large snapper, sheepshead and octopus. Ledges of two to four feet are typical, and depths generally run about 80 to 85 feet.

NO. 7 CRACKER RIDGE   44409.1   61959.0
Location: 12.5 nautical miles on a 70-degree course from the sea buoy at Ponce DeLeon Inlet.

A barge, dumpsters and several car transport trailers make up this artificial reef area. Depths are 80 feet to the bottom.

NO.8 PARTY GROUNDS   44390.2   61925.8
Location: 17.5 nautical miles on a 72-degree course from the sea buoy at Ponce DeLeon Inlet. This large reef area is marked with a north and south flag designated with the symbol PG on floats. The Loran coordinates are for the northwest section.

This area is good for grouper, snapper, cobia and lobster. The reef runs northeast and southwest, and has depths from 80 to 95 feet. Visibility ranges from 15 to 80 feet, but is usually at least 20 feet plus.

NO. 9 CULVERT REEF   44396.3   61972.8
Location: 12 nautical miles on an 80-degree course from the sea buoy at the Ponce DeLeon Inlet.

Tons of concrete culvert pipe is scattered on the ocean floor at a depth of 80 feet.

NO.10 EAST ELEVEN   44327.2   61927.3
Location: 19.2 nautical miles on a 90-degree course from the sea buoy at Ponce DeLeon Inlet.

This is a large reef with 8- to 10-foot ledges and depths from 66 to 75 feet. Large numbers of fish and some lobster are found here. There is also excellent shelling and tropical fish collecting. Visibility usually averages 20 to 30 feet.

NO.11 TURTLE MOUND   44257.4   61917.2
Location: Approximately 22.7 nautical miles on a 103-degree course from the sea buoy at the Ponce DeLeon (no markers).

This is another large reef with 10- to 12-foot ledges and depths of 60 to 70 feet. There are large numbers of fish, and visibility usually runs from 15 to 80 feet.

NO.12 FREIGHTER WRECK   44149.0   61795.8
Location: Approximately 50 miles offshore of Daytona on a 60-degree course from the sea buoy at Ponce DeLeon Inlet.

Feather duster   Ned DeLoach

This is a Liberty ship that is still loaded with jeeps and other war materials. It was torpedoed during WWII. Its remains are scattered over a large area of bottom, and many artifacts can be found. Spearfishing is great here, but the current is often strong (one to three knots).

NO.13 NORTH JETTY (Beach Dive)

Location: Runs approximately one mile into the Atlantic Ocean from the north shore of Ponce Inlet. Follow A1A south from highway 92 in Daytona Beach for ten miles to where it dead ends. Turn left and go to the beach. Then turn right and go one half mile and you'll see the granite jetty.

Large numbers of fish, stone crabs and some lobster inhabit this artificial reef. No spearfishing is allowed. Seek advice about tides and currents from local dive shops before diving here.

RAINBOW REEF (Not on Map)   44519.4   61959.1

Location: 16.8 nautical miles on a 32-degree course from the sea buoy at the Ponce DeLeon Inlet.

There are 400 metric tons of concrete culverts, measuring from three to 12 feet in diameter, scattered in the 2500-square-foot artificial reef site. Most of the material was put down in 1987. There is also the hull of a concrete houseboat in the area. The boat is 18 feet high and 32 feet long. Depths here vary from 73 to 78 feet.

SINKHOLE OFF FLAGLER (Not on Map)   44527.3   61495.6

Location: 3 nautical miles directly east of the Flagler Pier.

This cone-shaped ocean sinkhole starts at 80 feet and drops to 110 feet. There is little flow now coming from the vents at the bottom, but it was recorded that Spanish galleons once used the sink to replenish their freshwater supply. The area was used as a bombing site during WWII. Fragments of bombs are still found in the surrounding sand.

17
TITUSVILLE
MERRITT ISLAND - MELBOURNE

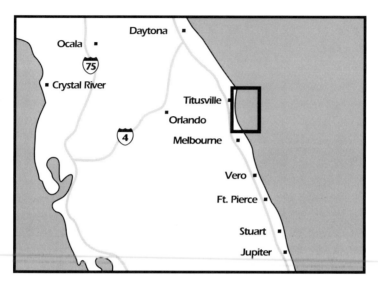

Offshore Diving Locations

1. *City of Vera Cruz*
2. *Leslie*
3. Dutch Wreck
4. Lead Wreck
5. *Damocles*
6. Pelican Flats

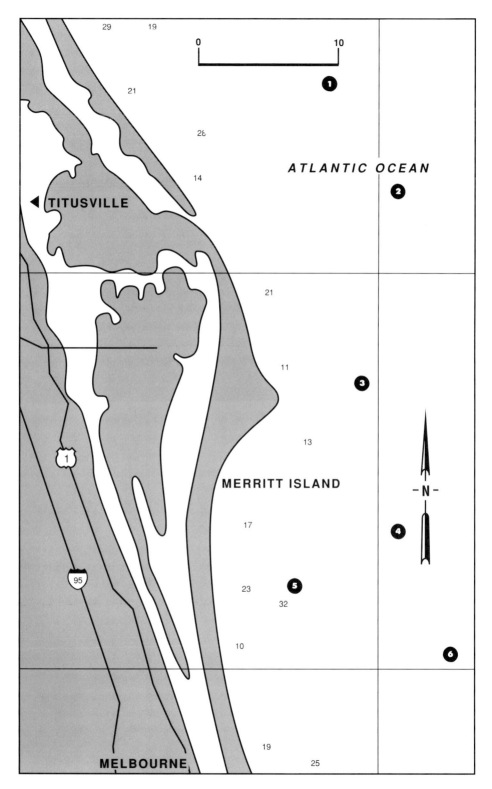

197

Area Information

The wreckage of several freighters sunk by torpedos during World War II and fields of limestone ridges typify the diving along the Space Coast. Fish and lobster are plentiful and spearfishing is the underwater activity of choice .

NO.1 *CITY OF VERA CRUZ*   44067.6   61910.2
Location: 25 miles NNW of Port Canaveral.

This 286-foot wooden steamship went down during a violent storm in 1880.as she followed the Florida coast on her way to Mexico. Sixty-eight crew members lost their lives. Today her broken wreckage covers nearly 100 yards in 80 feet of water. A great place to relic hunt.

NO.2 *LESLIE*   43972.1   60884.5
Location: Approximately 21 miles NNW of Port Canaveral.

A 200-foot coastal freighter was torpedoed in 1942. She rests in 80 feet of water.

NO.3 DUTCH WRECK   43931.9   61940.7
Location: Approximately 11 miles WNW of Port Canaveral.

The 400-foot freighter *Laertes*, heavily loaded with war materials, was torpedoed in 1942. The vessel has been salvaged on three occasions, but much still remains concealed in the heap of broken metal. She rises 28 feet off a 72-foot bottom.

Spadefish feeding above Atlantic wreck.   Mark Ullmann

NO.4 LEAD WRECK   43859.0   61926.3
Location: Approximately 17 miles east of Port Canaveral.
    This is another torpedo casuality from World War II. The freighter *Ocean Venus* was sunk in May of 1942. She rests in 70 feet of water with a relief of 36 feet.

NO. 5 *DAMOCLES*   43866.0   61920.0
Location: Approximately 10 miles ESE of Port Canaveral.
    This 148-foot coastal freighter was intentionally sunk as part of the Brevard Reef Site #1 in 1985. She sits upright in 85 feet of water. There are also concrete culverts in the area.

NO. 6 PELICAN FLATS   43780.7   61930.4
Location: Approximately 22 miles ESE of Port Canaveral.
    A large area of broken limestone ridges with depths to 80 feet. Best known as spearfishing and lobster grounds.

18

VERO BEACH

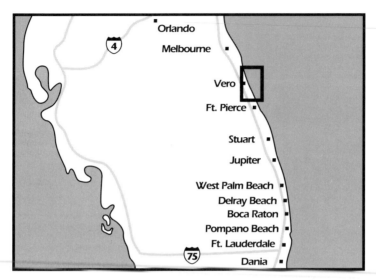

Offshore Diving Locations

1. Monster Hole
2. *El Capitana* and
   McLarty Museum
3. Wabasso Beach
4. Indian River Shores
5. Tracking Station
6. Jaycee and Conn
   Way Beaches

7. *Breconshire* Wreck
   (Boiler Wreck)
8. Humiston Beach
9. Riomar Reef
10. Cove Reef
11. Sandy Point Reef
12 Round Island
13. The Pines

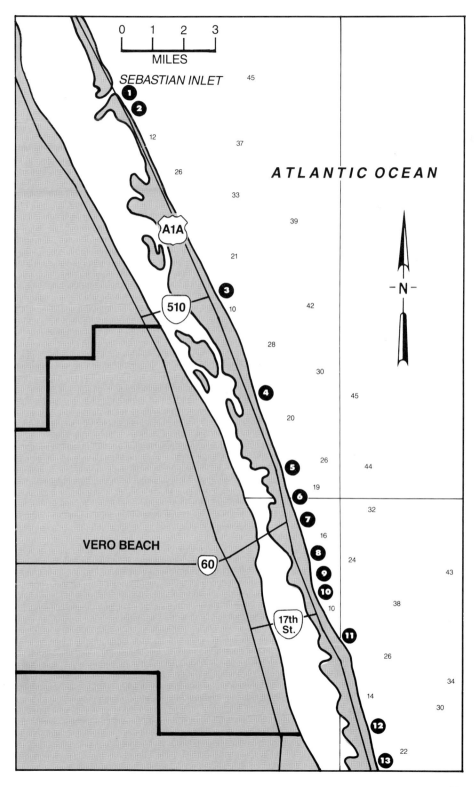

Area Information

The Vero Beach area offers a rarity for Florida—good beach diving. Although the clear waters of the Gulf Stream course well offshore, inshore visibility of up to 25 feet can be expected during periods of calm seas. Underwater sightseeing, lobstering,spearfishing and relic hunting are favorite diving activities.

A stellar attraction in the area is the scattered remains of cannon and ballast from the ill-fated 1715 Spanish Silver Plate Fleet. Violent hurricane winds forced ten treasure-laden ships onto inshore reefs. A pounding surf quickly broke the wooden vessels apart, scattering their valuable cargo of silver and gold. It was not until the late 50s that the wreck site was discovered and salvaged. The excavated riches produced one of the world's greatest underwater treasure finds. Today, the exciting story of the treasure fleet and the recovery of her precious cargo are told at the McLarty Museum, located in the Sebastian Inlet Recreational Area. An occasional cannon or ballast pile can be spotted by divers on local reefs and silver and gold coins are still being found on the beaches by beachcombers with metal detectors.

Offshore rock reefs run for miles parallel to the many beaches. There are four reef lines. The first (closest to shore) and second are low and partially covered with sand. A variety of tropicals, soft corals and sponges decorate the outcroppings. Approximately 200 to 400 feet offshore lie the third and fourth reefs, their depths varying from ten to 30 feet. Here, the underwater scenery is bolder, with high ledges often undercut by deep crevices. Rough seas not only limit water visibility, but create unsafe conditions for diving. However, when the conditions are right, you will discover a real underwater treat in the beach diving off Vero.

The Live Lobster Contest, held the first two weeks in August each year, has become a Florida tradition. Thousands of dollars in prizes are given in several categories. Make plans to try your luck with the monster lobster of Vero.

NO.1 MONSTER HOLE (Beach Dive)
Location: From the intersection of S-60 and A1A in Vero Beach, go north on A1A for 11 miles to the Sebastian Inlet State Park.

The hole is located about 35 yards off the jetties' end. It is 20 feet deep on the edge and drops to 30 feet in the center. Lobsters (big ones) are regularly yanked from the site. Be sure to trail a floating diver's flag.

NO.2 *EL CAPITANA* AND McLARTY MUSEUM (Beach Dive)
Location: From the intersection of SR 60 and A1A, go north on A1A for 11 miles,
OR
1 mile south of the Sebastian Inlet Bridge to the Sebastian Inlet State Recreational Area.

The *El Capitana* is one of the ten treasure galleons that were driven onto local reefs by violent weather in 1715. A few cannon and hundreds of ballast stones are scattered across the reef. The large reef area starts 20 feet offshore. Depths range from two to 20 feet. Because of the shallow depth, entries and exits can be hazardous during periods of rough surf.

Divers will be interested in visiting the state-run McLarty Museum which is located in the park. It is situated on the site of the salvage camp used by the survivors of the fleet. The exciting history of the fleet, its violent fate and the recovery of its fabulous wealth of gold and silver are recounted at the museum. It is open to the public Wednesday through Sunday from 9 a.m. to 5 p.m. Admission to the recreational area, including a visit to the museum, is 50 cents.

NO.3 WABASSO BEACH (Beach Dive)

Location: Go east from the intersection of SR 510 and A1A. This is 7.5 miles south from the Sebastian Inlet Bridge.

Wabasso Beach area is a large public park with plenty of parking and an easy boardwalk access to the beach. The reef starts 75 feet from the shore. Depths range from six to 15 feet. There are many ledges inhabited by an abundance of marine life.

NO.4 INDIAN RIVER SHORES (Beach Dive)

Location: From the intersection of SR 60 and A1A, go north for 5 miles. There are two access roads to the beach within 1 mile of each other. Diving is good off both the north and south access.

A long, wide section of limestone reef runs parallel to the beach, 150 feet offshore. The highest relief is on the seaside of the reef. Depths vary from 10 to 15 feet. Fish and lobster are abundant in the area. A second reef line is a good 100 yards swim seaward from the eastern edge of the first reef.

NO.5 TRACKING STATION (Beach Dive)

Location: Travel 2-1/2 miles north on A1A from the intersection of SR 60 and A1A. The beach access is located between a 7-11 and an Eckerd drug store.

Tracking Station is located at a large public beach with plenty of parking. The reef line starts about 200 feet off the beach. Depths range from 10 to 20 feet. The bottom is quite rocky, with piles of large stone slabs forming the reef.

NO.6 JAYCEE AND CONN WAY BEACHES (Beach Dive)

Location: Go 1 mile north from SR 60 on Ocean Drive.

This large public beach area is over one mile long. Beach entries can be made anywhere. There are four distinct reef lines starting 200 feet from shore. Depths average 20 feet. There are numerous ledges, eight to 10 feet high. Many are undercut with intersecting caves.

NO.7 *BRECONSHIRE* WRECK (BOILER WRECK) (Beach Dive)

Location: At the end of SR 60 (east), 150 yards east of the Ocean Grill Restaurant, a section of the boiler protrudes from the surface, making her exact location easy to spot.

This 200-foot steel ship went down on April 30, 1894. She rests in 15 to 20 feet of water. A conglomerate of marine growth covers her twisted remains. Natural reef ledges surround the site. Good lobstering and spearfishing are found here.

Florida lobster   Ned DeLoach

NO.8 HUMISTON BEACH (Beach Dive)

Location: From the end of SR 60, go south on Ocean Drive for 1/4 mile.

Humiston is a large public beach area with ample parking, restrooms and showers. Beach lifeguards can point out the better entry points for the reef. The first of four reef lines starts 100 feet from the shore. The last line is 400 feet off the beach.

NO.9 RIOMAR REEF (Beach Dive)

Location: From the 17th Street Bridge, go north on A1A for 1 mile. Turn east on Riomar Drive. Public parking is located at the end of the road. Public access is next to the Riomar Beach Club.

This is a large expanse of reef that sweeps in an arc from 10 to 500 feet off the beach. Depths vary from three to 20 feet. A few cannons from the Senora del Carmen a part of the1715 fleet are located in the area.

NO.10 COVE REEF

Location: From the intersection of S-60 and A1A, go south to the 17th Street Causeway. Turn left on 17th Street to the beach.

This is the same reef that runs the entire length of the Vero coast, but the shoreline dips inland for 3 miles creating the cove area. The high-profile reef, with ledges from 15 to 20 feet high, is 400 yards offshore. It can be reached by inflatable boats from the beach. This is a great spot for game fish.

NO.11 SANDY POINT REEF

Location: The reef can be reached only by boat due to the lack of public access from shore. It is situated 250 feet due east of the Moorings Development water tower.

Sandy Point consists of some nice ledges in 10 to 20 feet of water, which are considered excellent for spearing and lobstering. Cannons from the 1715 fleet are in the area.

NO.12 ROUND ISLAND (Beach Dive)

Location: 5 miles south of the 17th Street Bridge on A1A. Look for the entrance sign on the east side of the highway.

Round Island is another large public beach with easy access to the water. The reef is 150 feet offshore, in 15 feet of water. Good fish life and a few lobster inhabit the area.

NO.13 THE PINES (Beach Dive)

Location: From the 17th Street Bridge, go south on A1A for 7 miles. Look closely for a small gravel drive-through in a large cluster of pine trees. Or, go 3/4 of a mile north of the Bryn Mawr Camp Ground. Be careful not to get your car stuck in soft sand pockets on the road. Walk 300 feet north for water entry to close-in section of reef.

This is not a public beach, but it is open to the public. The reef starts 200 feet from the beach. Depths range from 10 to 20 feet. The Pines is one of the better sites for spearing and lobstering.

19

FORT PIERCE

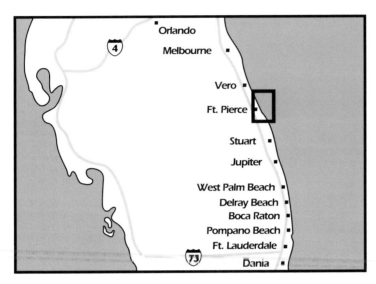

Offshore Diving Locations

1. Paddle-Wheeler Wreck
2. Pepper Park
3. Inlet Park
4. Jaycee Park
5. South Beach Rocks
6. Old South Bridge
7. The Fingers
8. Independence Reef
9. The Horse Shoe
10. *Amazon* (12A Wreck)
11. *Halsey* (Southeast Wreck)

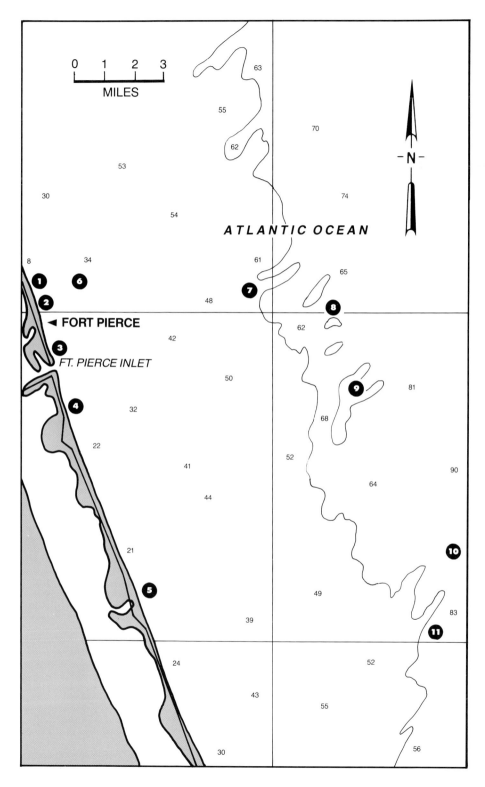

ATLANTIC OCEAN

◄ FORT PIERCE

FT. PIERCE INLET

Area Information

A Spanish treasure fleet, the Civil War, WWII and Mother Nature have all contributed to the adventure and beauty that is Ft. Pierce diving. For years the region has been known as a spearfishing and bull lobster haven, but today it is being acclaimed as Florida's up-and-coming recreational diving location.

NO.1 PADDLE-WHEELER WRECK (Beach Dive)
Location: .9 of a mile south of Bryn Mawr Camp Grounds entrance on A1A north.

The remains of a Civil War-era paddle-wheeler lie 100 yards off the beach. The ship rests on clean sand and can be seen easily from shore on calm days. Thousands of tropicals make their home among the broken ribs and boiler. Depths vary from 15 to 20 feet.

NO.2 PEPPER PARK (Beach Dive)
Location: On north A1A, approximately 2 miles north of the Ft. Pierce Inlet.

This is an all-time favorite of area divers. The formation begins as close as 100 yards from the beach, and is great for both snorkeling and scuba diving. Depths range from 15 to 30 feet. There is good lobstering, but spearfishing in the municipal park is prohibited. The park provides showers and excellent facilities.

NO.3 INLET PARK (Beach Dive)
Location: Just off A1A at the southern point of Ft. Pierce Inlet.

A large variety of rock formations and ledges stretch the length of the beach. Some are as close as 75 yards from shore. Depths range from 15 to 30 feet. This is a great spot for lobstering and fish watching. The remains of a small fishing boat can be found a quarter of a mile north of the Inlet. The facilities at Inlet Park are excellent and include picnic tables, showers and restrooms. The entrance fee is 50 cents.

NO.4 JAYCEE PARK (Beach Dive)
Location: On A1A, 1-1/2 miles south of the Inlet. Use the large water tower as a marker.

The ledges of the south beach area are a continuation of the chain of ledges that run from Ft. Pierce. There are three sets of ledges: at 15 to 20 feet; 25 to 30 feet; and 55 to 60 feet. Spearfishing is good and shelling is excellent. It is also a good location for lobster early in the season. Inshore diving ranges from 100 yards to 300 yards out from the beach, making Jaycee Park an ideal area for inflatables.

NO.5 SOUTH BEACH ROCKS (Beach Dive)
Location: On A1A south, at the first turn-off after the St. Lucie Nuclear Power Plant.

Marine life is plentiful and spearfishing is hot at South Beach Rocks. Barracuda are commonly sighted. While depths range from 10 to 18 feet, this area is not recommended for novices. Care should be taken to stay clear of the intakes of the nuclear power plant. Currents can be treacherous. Enjoy this dive, but use caution.

NO.6 OLD SOUTH BRIDGE

Location: 1 mile north of the Inlet.

The broken rubble from the Old South Bridge was dumped in 30 to 40 feet of water to create a fish haven. Bottom fish are plentiful, but lobsters are scarce. This is a good spot for fish watching. The fish haven has proven so popular that plans are underway to develop the area further.

NO.7 THE FINGERS

Location: Follow a 60-degree heading from the Ft. Pierce Inlet for about 12 miles.

The Fingers stretch for ten miles north and run at depths ranging from 45 to 66 feet. They are teeming with marine life, and seem to be a combination of beautiful Keys diving and true wilderness. Many snapper and grouper can be spotted roaming around the ledges. The beauty and marine life found here make this a must for the underwater photographer. Lobstering and spearfishing are good. Brightly colored tropicals are plentiful.

NO.8 INDEPENDENCE REEF

Location: Approximately 7 miles due east of the Ft. Pierce Inlet.

This spot is difficult to find; but, because of that, it harbors an abundance of marine life. Its ledges rise three to 10 feet off the bottom, and lie in 55 feet of water.

NO.9 THE HORSE SHOE

Location: Approximately 10 miles east of the Ft. Pierce Inlet.

Depths range from 55 to 65 feet. A wide variety of fish can be found at this site. Grouper, well over 15 pounds, are abundant. Tales of large lobster are commonly associated with this area.

NO.10 *AMAZON* (12A WRECK)   43206.1   61975.3

Location: 16 miles east-southeast of the Ft. Pierce Inlet.

This large freighter, torpedoed during WWII, lies scattered across the bottom in 100 feet of water. There are plenty of snapper and other game fish on the wreckage.

NO.11 *HALSEY* (SOUTHEAST WRECK)   43175.7   61987.0

Location: Approximately 13 miles east-southeast of the Ft. Pierce Inlet.

The freighter *Halsey* was torpedoed during WWII. She still remains fairly well intact, but broken into three sections. The stern section, which still supports part of the superstructure, rests upright, as does the bow. The mid-section is upside-down. This is one of Florida's best wreck dives from the WWII era.

20

STUART – JUPITER

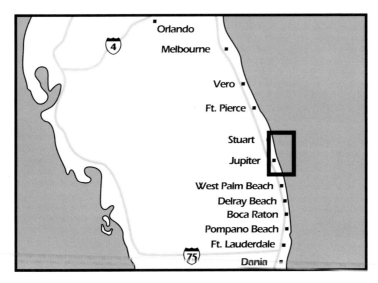

Offshore Diving Locations

1. House of Refuge-
   Old Schooner Wreck
2. Bathtub Reef
3. North Jetty Reef-
   St. Lucie Inlet
4. Toilet Bowl Reef
5. Six Mile Reef
6. LST-St. Lucie Inlet
7. Tire Reef
8. Pecks Lake and
   Kingfish Hole
9. *Gulf Pride*
10. Blowing Rocks
11. Coral Cove Park
12. South Jupiter Island
    Bridge
13. Carlin Park
14. The Ranch
15. Grouper Hole
16. Barrow Reef
17. Reformation Reef
18. Jupiter High Ledge
19. Rio Jobe Reef
20. Juno Ball Ledge
21. The Cave
22. USS *Rankin* / Evinrude
    Memorial Reef
23. *Esso Bonaire III*

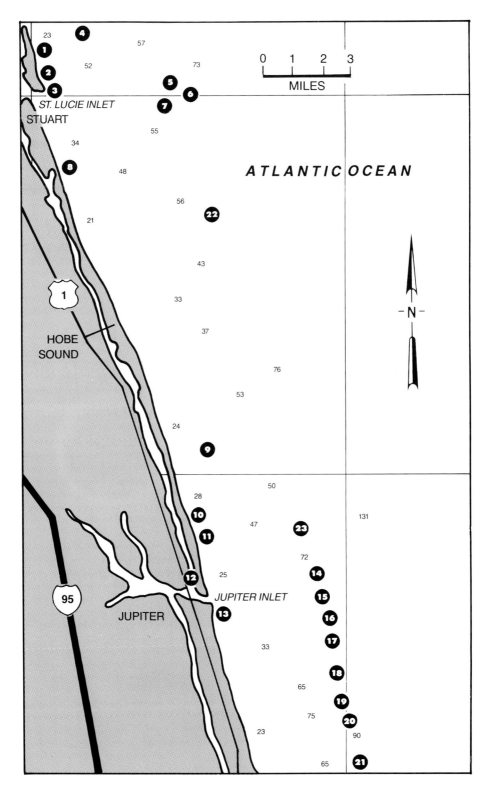

23

④

57

①

52

73

②

5

③

6

ST. LUCIE INLET

7

STUART

55

34

⑧

48

ATLANTIC OCEAN

56

㉒

21

43

🛡1

33

HOBE
SOUND

37

76

53

24

⑨

50

28

131

⑩

47

㉓

⑪

72

12

25

⑭

JUPITER INLET

⑮

🛡95

13

⑯

JUPITER

⑰

33

⑱

65

⑲

75

⑳

90

23

㉑

65

0   1   2   3

MILES

– N –

Area Information

The Atlantic reefs from Jupiter through Stuart are generally continuations of the rock ledges that parallel the West Palm coastline. However, there are two major differences: water visibility is less here, and there are fewer divers exploring these waters.

The Florida peninsula inclines gradually westward as it proceeds north of West Palm. Reefs following the contour of the land are abandoned by the Gulf Stream's clear water as it continues its northerly flow. Murky outflow from rivers and water periodically released from Lake Okeechobee also limit visibility. Summer diving is considered best. Then, calm and clearer waters usually prevail. Diving during winter months is more demanding but offers a greater variety of fish and more lobster.

Because they are dived less, the rock reefs of the area are not as easily identified as those to the south. They are, however, quite large and extensive, sometimes running continuously for miles. Prominent reefs begin three miles offshore in 60 to 65 feet of water. The outer ledges are impressive but deep. They are in 120 to 160 feet of water. Divers accustomed to the shallower reefs in southern Palm Beach County find the northern reefs more sparse in terms of marine life such as sea fans and gorgonians. They more than redeem themselves with their abundance of larger species (grouper, amberjack, turtles, and, of course, lobster).

NO.1 HOUSE OF REFUGE — OLD SCHOONER WRECK (Beach Dive)
Location: Just off A1A at the south end of Hutchinson Island.

The House of Refuge is an old lifesaving station, the last of its kind. Now a museum, it is well worth a visit to see the displays and the turtles that are raised there. The turtles are kept until they are big enough to have an increased chance of survival; then released to the oceans.

The ribs of an old schooner can be found by swimming 100 yards straight off the south end of the concrete block wall in front of the museum. Sand dollars are sometimes found on the sandy flats that lead to the schooner.

NO.2 BATHTUB REEF (Beach Dive)
Location: On Hutchinson Island, travel south on MacArthur Blvd. to the public beach at the end of the road.

This is a popular snorkeling site. The best diving is at high tide on the shallow rock reefs that parallel the beach.

NO.3 NORTH JETTY REEF / ST. LUCIE INLET (Beach Dive)
Location: From the tip of the north rock jetty at the St. Lucie Inlet, head due north toward the Bathtub Reef area.

This is a good drift dive over rock ledges in four to 20 feet of water. Be sure to look for large snook hiding under the reefs. Early in the season, many lobster of legal size and a few nice size stone crabs are in residence. Make sure they are in season, though, as the Florida Marine Patrol diligently protects this area.

NO.4 TOILET BOWL REEF   43107.3   62013.1
Location: From the St. Lucie Inlet, travel 3-1/2 miles on a 10-degree course, or about

2 miles due east of the House of Refuge.

The area is aptly named because this artificial reef was formed by the dumping of old toilets. The depth is 60 feet. It will certainly be a memorable dive and it might be possible to "flush" out a fish or lobster. There are also a couple of barges on the site.

NO.5 SIX MILE REEF

Location: From the St. Lucie Inlet, travel 6 miles on an 80-degree course.

This is a nice ledge in 75 to 80 feet of water. Snapper and grouper are found in the area. A current is usually running to the north.

NO.6 LST — ST. LUCIE INLET   43090.5   61992.9

Location: From the St. Lucie Inlet, travel 6-1/2 miles due east.

A landing craft and sand dredge were sunk in this artificial site known as the Capt. Al Sirotkin Reef in 1982. The LST rests in an upright position in 85 feet of water. It was stripped before being sunk.

NO.7 TIRE REEF   43062.9   62003.0

Location: From the St. Lucie Inlet, travel 4 miles on a 115-degree course.

This well-established artificial reef project, named the Dr. Edgar Ernst Reef, was started in 1971. The old tires form excellent habitats for marine life. There are also barges and bus bodies on the site. The frames of the buses have deteriorated rapidly. The depth is 58 feet. Two additional sets of numbers for materials in the site are: **43063.7 62002.4** and **43066.3 62002.7**.

NO.8 PECKS LAKE AND KINGFISH HOLE

Location: 3 miles south of the St. Lucie Inlet, adjacent to Pecks Lake in the Intracoastal Waterway.

The Kingfish Hole is about 900 yards offshore in 34 feet of water. Lobster abound in this area. Depths vary from 2 to 20 feet. This is an excellent place to free-dive when the weather is clear.

NO.9 *GULF PRIDE*   14351.9   62026.2

Location: 4 miles due north of the Jupiter Inlet.

A victim of a collision during WWII, the Gulfland tanker burned and split into two sections. Today, the wreckage supports a huge variety of marine life, ranging from swarming schools of baitfish to large barracuda and grouper. Because of its close proximity to the beach (about one mile offshore), it should only be dived during periods of calm seas when there is clear water. She rests in 40 feet of water.

NO.10 BLOWING ROCKS (Beach Dive)

Location: 2 miles north of Jupiter Inlet on A1A, at the beachfront on Jupiter Island. The area is called Blowing Rocks Preserve. The reefs begin just about where the condos end and continue north to Hobe Sound.

A very craggy, rocky shoreline (footwear is recommended) extends below the surface of the water in the form of holes and ledges that are favorite haunts of snook and other schooling fish. This area can be rough to dive when the surf is up, but on calm days it is an excellent spot. There is plenty of parking available.

NO.11 CORAL COVE PARK (Beach Dive)
Location: 1 mile north on A1A from Jupiter Inlet.

This is a popular park with plenty of parking. The sandy beach has rock formations extending out to a depth of 15 to 20 feet. Best at high tide. Some lobstering and a variety of tropicals.

NO.12 SOUTH JUPITER ISLAND BRIDGE (Beach Dive)
Location: On the south end of Jupiter Island, take the bridge over the Intracoastal just north of the Coast Guard Station. The road is marked as A1A or SR 707. Go down the dirt road on the bridge's left bank.

Dive at high, slack tide for the best visibility and least current. Stay out of the Intracoastal channel and fly a dive flag. The depth near the bridge footings just off the beach is approximately 25 feet. There are many tropicals here, with some stone crabs and lobster early in the season.

NO.13 CARLIN PARK (Beach Dive)
Location: A county park on the beach just south of the Jupiter Hilton.

A good snorkeling and diving spot with crumbled rocks extending out to about 15 feet of depth. The surf can be rough here, but on calm days the waters clear quickly. Many tropicals adorn the rocks. The park provides excellent parking and picnic facilities.

NO.14 THE RANCH
Location: Approximately 1/2 mile north of the Jupiter Inlet.

This is a colorful dive with depths from 70 to 82 feet. It is usually great for lobster, which hide in the many cracks and crevices. This is a good site for underwater photography of the many tropicals and sea fans that flourish here.

NO.15 GROUPER HOLE
Location: 1-1/2 miles due east of the Jupiter Inlet on the first reef line.

The reef parallels the shoreline and runs north. Depths range from 70 to 80 feet. Surrounding waters can be very clear on calm days. Deep, tunneled ledges about five feet in height make ideal havens for grouper during their annual winter migration. Some areas of the reef have back openings or "blow holes" large enough for divers to swim through.

NO.16 BARROW REEF
Location: 1/2 mile south of Jupiter Inlet.

This site is teeming with large angelfish, tropicals and a colorful array of sea fans. Large schools of fish are always present. This is a good place to start lobstering as you work your way down the reef line. Depths vary from 60 to 72 feet.

NO.17 REFORMATION REEF
Location: 1 mile south of Jupiter Inlet and 2 miles offshore.

Large schools of yellowtail snapper and amberjack are often found in this section of the reef. Turtles also frequent the area. Depths are from 55 to 80 feet.

NO.18 JUPITER HIGH LEDGE
Location: 1-1/2 miles south of the Jupiter Inlet.

A spectacular dive for more experienced divers. This is the highest relief of any Palm Beach County reef, 90 feet in the sand at the base and 65 to 70 feet at the top of the reef. A huge, stepped ledge includes crumbled boulders and crevices galore. This ledge lends itself to drift diving. Due to its sheer mass and myriad hiding places, the reef supports an abundance of marine life in many sizes, shapes and species. Schooling snapper, amberjack and grunt can be so thick that they occasionally obscure the reef from view. Grouper, lobster, moray eels and tropicals lurk about the shadowy holes and crevices.

NO.19 RIO JOBE REEF
Location: 2 miles south of Jupiter Inlet.

In this exciting section of the reef line, a great variety of rays make their home. You might have the fortune to sight a manta or eagle ray soaring over the reef. Depths range from 60 to 75 feet.

NO.20 JUNO BALL LEDGE
Location: 4 miles south of Jupiter Inlet and 3 miles offshore.

This reef has a spectacular drop-off which starts in 65 feet of water and plummets to 85 feet. This ledge is ideal for drift diving, and includes a flagstone design of corals on the wall. You can float down the wall, drifting with game fish and sea turtles.

NO.21 THE CAVE
Location: 5 miles south of the Jupiter Inlet and approximately 3-1/2 to 4 miles offshore.

Among the most awesome sights available to the underwater observer, the top of this reef is in 120 feet of water, its sides plunging downward to 160 feet. The cave, actually better described as a tunnel, cuts into a corner of this drop-off and elbows right back out again. The tunnel is big enough to drive two semis, side by side, all the way through. It is 140 feet to the cave floor. The inside of the tunnel is full of ledges where 300-pound jewfish, 45-pound snappers and 50-pound grouper park themselves. Lobster here are big enough to scare you. No lights are necessary. Once you enter the tunnel, you can see the huge exit. The length of the tunnel is approximately 60 to 80 feet. Schools of fish such as mackerel, tuna and amberjack whisk past the opening, waiting to catch glimpses of divers. **This dive is for experienced divers only.** Strong currents and sharks are common.

NO.22 USS *RANKIN*/EVINRUDE MEMORIAL REEF   14373.1   61986.7
Location: 6 miles southeast of the St. Lucie Inlet.

The *Rankin* is the largest ship intentionally sunk as an artificial reef on the east coast. The huge 459-foot amphibious assault ship was put down during the summer of 1988. She rests on her side in 120 feet of water. The main deck is 80 feet down. What a dive!

NO.23 *ESSO BONAIRE III*   14351.3   62006.5
Location: 4 miles east-northeast of the Jupiter Inlet.

This 147-foot harbor tanker was sunk in 1989. The steel vessel was constructed in 1926. She lies in 90 feet of water.

21

WEST PALM BEACH

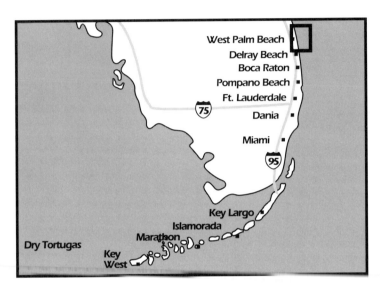

Offshore Diving Locations

1. Valley Reef
2. Koller's Reef
3. *Mizpah*
4. *Amarilys*
5. Spearman's Barge
6. Double Ledges
7. Artificial Reef Site 2
   (*Owens*, Rolls Royce, Barge)
8. Tri-County

9. Rock Piles
10. Breaker's Reef
11. The Trench
12. Cable Crossing
13. Ron's Reef
14. Midway
15. Paul's Reef
16. Horseshoe Reef

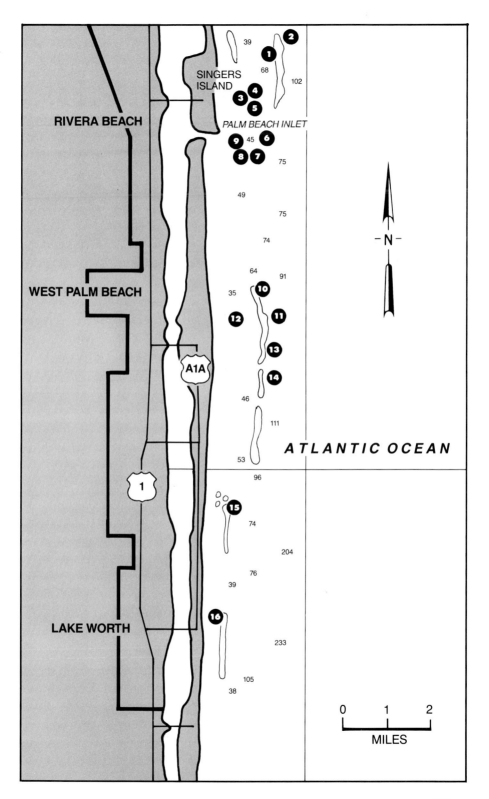

RIVERA BEACH

SINGERS ISLAND

PALM BEACH INLET

WEST PALM BEACH

A1A

ATLANTIC OCEAN

1

LAKE WORTH

-N-

0   1   2
MILES

Area Information

Diving the reefs along the shores of the Palm Beaches provides a sensational sense of exploration and discovery. Local divers refer to themselves as "Palm Beach Divers" because of their pride in the rugged, untamed beauty that envelops the reefs. Palm Beach diving has an advantage over all other diving locations because of one main factor—the Gulf Stream. This warm current swings closer to shore here than anywhere else along the coast of Florida. The result is warm water diving year-round and unbelievably clear, clean water that is continually feeding and refreshing the reefs, keeping them well-stocked with beautiful tropicals, lobster, huge game fish, sea fans and corals. The gentle currents let divers experience drift diving, and enable them to explore miles of beautiful reefs with little swimming effort! One of the great conveniences of Palm Beach diving is that all the reefs, ranging in depth from 30 to over 100 feet, are well within two miles of land and close to the inlet. The Palm Beach Inlet is wide and deep, making boating quite easy and safe even for beginners.

The diving facilities of the Palm Beaches offer vacationing divers everything they could ask, making trips exciting, convenient and diverse. Reef trips for single or two-tank dives are available. The diving operations are complete facilities, supplying guide services, sales, rentals, instruction, repairs, photo services and air. Marinas and boat ramps are located only minutes from the Palm Beach Inlet. The Palm Beaches is a vacation resort offering a wide variety of motels, fine restaurants and beaches.

A year-round look at Palm Beach diving shows that the summer months of May through September offer the calmest seas, ranging anywhere from flat as a lake to two to three feet, depending on the South Florida weather picture. From January through March, temperatures from the low to mid-60s can be expected at night, with 75 to 85 degrees as daytime highs. Charter boats still run trips when the seas are five to six feet for the hardy divers. The possibility of flat seas, however, exists throughout the year. For those on vacation, good diving can be guaranteed some time, some place, during the week. April weather starts changing again to summer conditions.

There is always a great abundance of marine life for the avid spearfisherman, tropical fish collector, shell collector, photographer or sightseer. Visibility is normally 50 feet plus, though it is not uncommon to get 80- to 100-foot visibility.

The general attitude of the boat captains in the Palm Beach area is one of cooperation and safety-mindedness toward all divers coming into the area. Feel free to stop by at any of the dive shops that charter for information on local diving areas and conditions. It should be a must on your agenda.

NO.1 VALLEY REEF

Location: Approximately 2-1/2 miles from shore and a boat ride of 15 minutes from Palm Beach Inlet.

This reef is great for a "quickie" dive because it is so close to the Inlet. It runs north and south, making it another terrific drift dive possibility. There is a ledge on both sides of the diver, creating a valley effect down the middle. The depth in the valley is 75 feet, and the walls of the ledges vary between three and ten feet in height. A diver can drift this reef for a full 40 minutes bottom time and never run out of reef. Shelling is great, and the reef is abundant with game fish and tropicals.

NO.2 KOLLER'S REEF
Location: Approximately 3 miles northeast of Palm Beach Inlet. Boat ride is about 15 minutes.

Depths at Koller's Reef range from 75 to 85 feet. If you wish, you can drift dive the area, following the reef out to 125 feet. This is a very winding reef which meanders north and south. Sea turtles are quite commonly sighted, and they have no fear of divers, often coming in close to have their pictures taken. Spearfishing is great in this area, as it is on most of the northern reefs, which are heavily populated with red snapper and grouper. These rock ledges are covered with corals; and, in combination with the many colorful tropicals, lend intricate beauty to the reefs.

NO.3 *MIZPAH*   14332.9   62027.8
Location: Approximately 1-1/2 miles from shore and about a ten-minute boat ride from Palm Beach Inlet.

The *Mizpah* is an old Greek luxury liner, 185 feet long and completely intact, lying in 90 feet of water. The vessel was sunk in 1968 to serve as an artificial reef and fish preserve. Therefore, spearfishing, shell collecting and lobstering are prohibited. Lying next to the *Mizpah* is a 165-foot patrol craft. The two ships rest

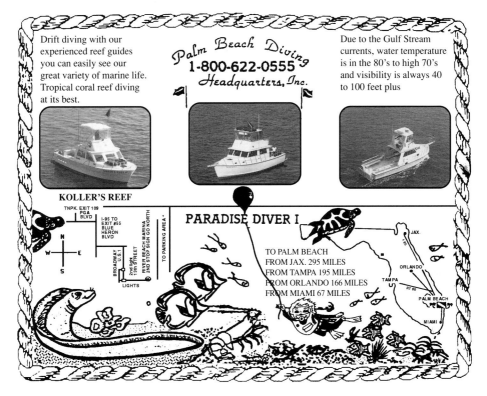

Drift diving with our experienced reef guides you can easily see our great variety of marine life. Tropical coral reef diving at its best.

Palm Beach Diving
1-800-622-0555
Headquarters, Inc.

Due to the Gulf Stream currents, water temperature is in the 80's to high 70's and visibility is always 40 to 100 feet plus

KOLLER'S REEF

TNPK. EXIT 109
PGA BLVD
I-95 TO EXIT #55 BLUE HERON BLVD
BROADWAY U.S.1
2nd light 13th STREET
RIVER BEACH MARINA 2ND STOP SIGN GO NORTH
TO PARKING AREA *
LIGHTS

PARADISE DIVER I

TO PALM BEACH
FROM JAX. 295 MILES
FROM TAMPA 195 MILES
FROM ORLANDO 166 MILES
FROM MIAMI 67 MILES

JAX.
ORLANDO
TAMPA
RT 60
PALM BEACH
MIAMI

on the bottom, side by side, right sides up. This dive site makes getting into wreck diving easy, safe and fun. No lights are needed throughout most of the wreck because it is well exposed to light. All the doors have been removed and there are no loose cables. This has been done for the safety of divers. The wreck houses tons of fish and corals. The stern section of the patrol craft lies approximately 100 yards northeast of the *Mizpah*. Both sections of the wrecks abound with marine life as well as an impressive collection of the spiny oysters. Nearby are the broken remains of a 65-foot wooden tug in 80 feet of water.

NO.4 *AMARILYS*   14331.1   62027.8

Location: This wreck is located just 300 yards northeast of the Mizpah, a short distance from the Palm Beach Inlet.

The *Amarilys* is a 441-foot steel freighter. Only its hull and lower deck remain. Because the wreck lies out in open sand, it has become a home for all the fish and marine life of the vicinity.

NO.5 SPEARMAN'S BARGE

Location: East of the Mizpah. 2 miles from shore and a ten-minute boat ride from Palm Beach Inlet.

When people think of Palm Beach wreck diving, this little barge doesn't always come to mind because of the *Mizpah* and the *Amarilys*. As a result, large game fish constantly surround the barge and never hesitate to greet a diver. Everything on the barge is huge: there are morays, tropicals, 50- to 100-pound grouper and, occasionally, a 300- to 400-pound jewfish. Shelling is quite good and a few rare specimens have been found. The barge is also full of spiny oysters. Depths range from 65 to 75 feet. Unlike the other wreck sites, the barge is surrounded by natural reefs. The barge was sunk in 1978.

NO.6 DOUBLE LEDGES
Location: Approximately 1-1/2 miles straight out from the Palm Beach Inlet, ten minutes from the dock.

This exceptionally large ledge runs north and south and starts in 70 feet of water at the top of the reef. The first drop-off is to 80 feet; then it drops off again to 90 feet. Deepwater game fish, such as sailfish and schools of mackerel, are sighted as well as the largest and most docile loggerhead turtles found in the Palm Beach area. Drifting along you'll find yourself atop a yellow blanket made up of thousands of porkfish. There are several such "Porkfish Holes" along this reef. Jewfish, pygmy angels, and lobster are common. Shell collecting is excellent.

NO.7 ARTIFICIAL REEF SITE 2
(*Owens*, Rolls Royce, Barge)   14331.1   62028.7
Location: About 1 mile southeast of the Palm Beach Inlet.

Since its construction in 1985, this area has quickly become one of the most popular dive sites of the Palm Beaches. Depths are from 80 to 90 feet.

Owens
Sunk in December of 1985, the *Owens* is the largest addition at Site 2. Located just 5 minutes from the mouth of the Inlet in 85 feet of water, the 125-foot freighter rests upright. The diver can drop down into huge cargo holds and exit the hull through large holes that were blown out by charges to sink the ship. Swim through the crew's quarters and look out portholes to see just how much marine life the ship has accumulated already. Algae and small sponges, corals and tiny gorgonians are beginning to adhere to the wreck. The barracudas have schooled around the wreck since the second day of her sinking. Grouper to 30 pounds are seen on every dive. From her bow you can see both the Rolls Royce and the Barge.

Rolls Royce
Where else but in Palm Beach would someone donate a mint condition Silver Cloud Rolls Royce to kick off the county's Artificial Reef Committee? In the beginning, it was truly unique to see on the bottom a highly polished Rolls, complete with the hood ornament. Vandals have since seized the "Lady" and the polish soon disappeared. Tiny lobster have taken up residence under the back seat, and porkfish now school under the hood.

The Barge
The Barge has been down since April 1985 and is loaded with tropicals. Snapper congregate around the bow while nurse sharks and grouper hide under her prow. A group of divers can dive the barge, drift back and see the Rolls, then swim over and explore the *Owens*—all in one fantastic dive.

NO. 8 TRI-COUNTY   14330.2   62031.8

Location: 3/4 mile south-southeast of the Palm Beach Inlet. .6 nautical mile offshore.

This is an artificial reef site 200 x 100 feet in size that is made up of 1800 tons of concrete culverts. The area has a 12-foot profile from a depth of 60 feet.

NO.9 ROCK PILES   14330.7   62033.0

Location: A large spoil area approximately half a mile in diameter, lying a short distance southeast of the south jetty of the Palm Beach Inlet.

These large rock piles consist mainly of rubble from the dredging of the Inlet and turning basin. Depths range from 50 to 70 feet. The many rock piles are spaced 10 to 50 feet apart. The location of the Rock Piles protects them from winds that might disturb other diving areas. The area has become a home to many forms of marine life.

NO.10 BREAKER'S REEF

Location: Approximately a half-hour boat ride from the Palm Beach Inlet and 1 mile from shore, straight out from the Breaker's Hotel.

Breaker's Reef is very long and runs north and south, twisting occasionally. Depths range from 45 to 60 feet at the bottom. Because it is extremely craggy, this reef is loaded with fish of all kinds. It also makes a splendid home for the Florida spiny lobster. Because the reef is so long and diversified, the diver will find he is on the verge of discovery with every bend he turns. Huge schools of harmless, friendly barracuda are one of the great sights, as is the abundance of sea turtles.

NO.11 THE TRENCH

Location: Approximately 1 mile south of the Breaker's Reef.

The Trench is a man-made cut lying in an east-west direction through a natural reef. One hundred yards in length, the trench is about 20 feet across and has a ten-foot drop on both sides. A virtual aquarium in 45 to 55 feet of water, the Trench is abundant with spadefish and porkfish. If the timing is right, you may come up with a bagful of lobster. On every dive, giant turtles are spotted and an occasional nurse shark can be found sleeping under a ledge. The dive is ideal for both novice and experienced divers.

NO.12 CABLE CROSSING

Location: Approximately 1/4 mile offshore, following a half-hour boat ride south from the Palm Beach Inlet.

Having decided to protect this area so that all divers could share in its beauty, the Palm Beach divers have left sea fans and corals to grow unmolested. As a result, divers will witness a most colorful reef full of juvenile tropicals, purple sea fans, meandering rock ledges full of corals and sponges, many rays, and, occasionally, some huge tarpon wandering through. Because of its shallow depth of only 30 feet, there is never any current and one can enjoy this area with only snorkel gear.

NO.13 RON'S REEF
Location: 2 miles south of the Breaker's Reef.

There is a natural channel in this reef section that attracts fish from the outer reefs. There fish take up residence in the large overhanging ledges. A school of spadefish and grass eels is found in the sand just north of the ledge. Swimming west, you will find a fish "cleaning station" that barracuda have been attending for years.

NO.14 MIDWAY
Location: A 35-minute boat ride south from the Inlet. About 3/4 of a mile offshore.

This stretch of reef is about a mile long with a depth of 45 feet on top and 60 feet in the sand. Fish are everywhere. Schooling fish and tropicals sweep through the coral arches and caverns. Morays, nurse shark, and rays are commonly sighted. Another great feature here is the almost total absence of current. This makes it a great spot for the novice diver and underwater photographer.

NO.15 PAUL'S REEF
Location: 4 miles south of Breaker's Reef.

This reef is the north end of a long reef formation, with a depth of 45 feet at the top and 60 feet in the sand. The area is noted for the large blue and green parrotfish that seem to be found around every turn. The reef's walls are overgrown with soft corals, sea fans and hard corals. Many expose their feather-like polyps to the soft currents, even during daylight hours. At the north end is a coral arch through the reef that leads into a natural fishbowl. Large holes honeycomb the reef, providing an abundance of hiding places for large snapper and grouper. Drift diving is excellent over the white sand bottom surrounding the reef, where sand dollars, fighting conchs and helmet shells can be found.

NO.16 HORSESHOE REEF
Location: The northern tip of an extensive reef formation that parallels the coast in a southerly direction. Horseshoe Reef lies about 1 mile northeast of the Lake Worth pier.

Depths on Horseshoe Reef vary from 40 feet at the top of the reef to 55 feet in the sand. The inner ledge is a ten-foot vertical drop undercut in many areas by small caves and ledges. The reef is covered with a beautiful array of sponges and soft corals. Rays, barracuda and amberjacks are common. Many brightly colored tropicals can be seen darting in and out of the undercuts.

22

LAKE WORTH – DELRAY BEACH

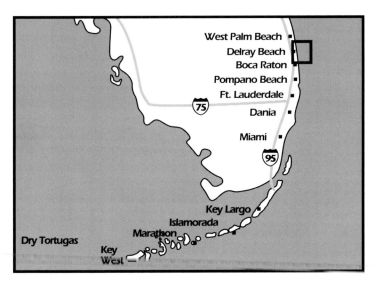

Offshore Diving Locations

1. Football Field
2. Hogfish Reef
3. Feather Beds and Barge
4. Loggerhead Reef
5. Lynn's Reef
6. Briney Breeze
7. Boynton Falls
8. Andy's Patches
9. Gulf Stream Ledge
10. Grouper Hole
11. Budweiser Bar Wreck
12. Genesis Reef
13. Delray Ledge
14. Mountains and Valleys
15. Delray Wreck

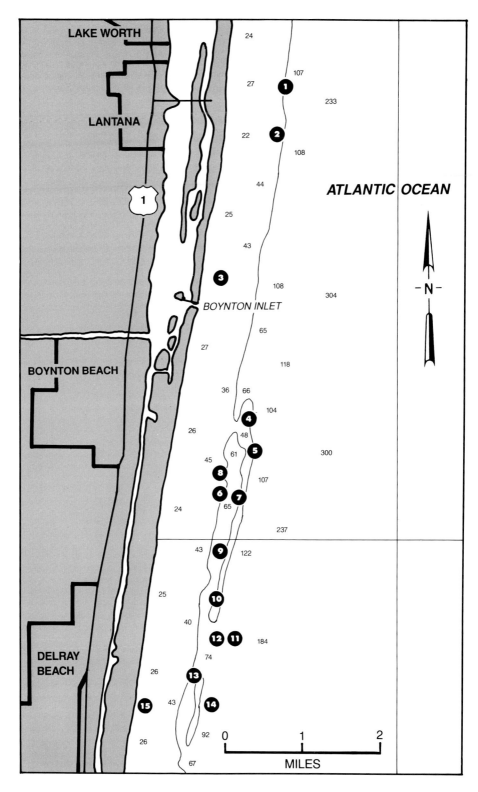

LAKE WORTH

LANTANA

BOYNTON BEACH

DELRAY BEACH

BOYNTON INLET

ATLANTIC OCEAN

- N -

24

107

27

233

22

108

44

25

43

108

304

65

27

118

36   66

104

26

48

300

45   61

107

24   65

237

43   122

25

40

184

74

26

92

43

26

67

0   1   2

MILES

NO.1 FOOTBALL FIELD
Location: Approximately 3 miles north of Boynton Inlet.

Two low-profile reef lines run parallel for 100 yards before closing at the ends. Between the reefs is a field of crushed coral. Depths vary from 40 to 60 feet. The rock reef is covered with soft coral and sea fans.

NO.2 HOGFISH REEF
Location: 2-1/2 miles north of the Boynton Inlet.

The rolling contoured bottom is covered with large growths of grass that shelter an abundance of tropicals and lobster. Depths vary from 45 to 75 feet. Low-profile patch reefs are located just to the west, where depths range around 35 feet. The small reef heads support a community of soft corals and basket sponges. Hogfish are uncommonly plentiful. This is an excellent area for spotting large sea turtles.

NO.3 FEATHER BEDS AND BARGE
Location: 1/2 mile north of Boynton Inlet.

This shallow 25-foot dive is excellent for check-out dives and tropical fish collecting. Baitfish are prolific here. Local fishermen use the area to catch their bait. A 20-x 10-foot concrete barge is 50 yards southeast. The wreck rises 5 feet off the bottom. This is a hot site for lobster during the season. It is best to dive here during an incoming tide.

NO.4 LOGGERHEAD REEF
Location: One mile south of Boynton Inlet.

This is the most shallow dive on a good reef system in the area. The reef is over 300 yards long. Its inshore edge starts in 65 feet of water and gently rolls to a crest at 40 feet. The ocean side depth is 80 feet in the sand. There are no sharp ledges. Soft and hard corals cover the limestone bank. In the late spring and early summer, Loggerhead turtles are seen on every dive. Because it is close to the Inlet, boat traffic is heavy.

NO.5 LYNN'S REEF
Location: Just over 1 mile south of the Boynton Inlet.

Depths vary from 75 feet in the sand to 40 feet on top of the reef. Atlantic sea life is spectacular here. A great variety of tropicals and game fish frequent the site. This is another area with heavy weekend boat traffic. Fly your dive flag!

NO.6 BRINEY BREEZE
Location: 2-1/2 miles south of the Boynton Inlet, just offshore from the neighborhood of Briney Breeze.

This low-profile reef has many cracks and crevices. The reef's top is 45 to 50 feet underwater. Depths to the sand are 65 feet. Morays and nurse sharks can be found under the ledges.

NO.7 BOYNTON FALLS
Location: Seaward from Briney Breeze.

Ned DeLoach

One of the deepest drops on the outside of the reef. Depths are from 70 to 80 feet. There are several large overhangs and deep cuts. It is best to drift the area because of a strong northern current.

NO.8 ANDY'S PATCHES
Location: 3 miles south of the Boynton Inlet.

Just inside the main reef system is a series of low-profile patch reefs. The rock islets extend for 1/4 mile and run about 100 yards inshore. Most are 10 to 20 yards apart. Each miniature reef is alive with marine life.

NO.9 GULF STREAM LEDGE
Location: 4 miles south of the Boynton Inlet.

This dramatic 15-foot-high ledge runs for over 200 yards down the inside of the reef. The bank is 100 yards wide. On the outside, reef fingers extend 100 feet seaward, losing their profile as they go. It is a 60-foot dive to the reef's top. The sand areas between the projections are 80 to 100 feet deep. Long, low caves undercut the north side. Sea whips, sea fans and corals cover the walls.

NO.10 GROUPER HOLE
Location: 4 miles south of Boynton Inlet on the south end of the fingers.

For years grouper have made this small cave their home. When the fish see divers come over the last finger, they quickly dodge into the cave entrance. The interior is only 6 to 8 feet wide. The depth is 60 feet.

NO.11 BUDWEISER BAR WRECK   14300.5   62065.0
Location: 4-1/2 miles south of the Boynton Inlet. One nautical mile offshore.

With the financial help of Budweiser, this 167-foot coastal freighter was placed down in July 1987 as part of the Palm Beach Artificial Reef Program. It rests in 87 feet of water. The deck is 70 feet under and has a large open section that can be easily entered. Davits, funnels and the propeller were left intact. This is an excellent dive.

NO. 12 GENESIS REEF   14301.2   62066.0
Location; Just inshore of the Budweiser Bar Wreck.

This is a 200-by 100-foot area of concrete culverts. The depth is 70 feet. The material has a 10-foot profile.

NO.13 DELRAY LEDGE
Location: Starting 5-1/2 miles south of the Boynton Inlet.

Dramatic underwater scenery is at its best here. The almost vertical ledge is more pronounced than any other on the "third reef." Large sections have broken from the main reef, forming narrow tunnels and passages. Caves and overhangs filled with sea life are waiting to be explored. The reef's crown supports an outstanding growth of sponges and soft corals. Snapper, grouper, grunts and tropicals are common. An occasional nurse shark can be spotted hiding under the protection of an overhang. Depths range from 50 to 65 feet in the sand.

NO.14 MOUNTAINS AND VALLEYS
Location: On the outside of Delray Ledge.

An excellent 1/4-to 1/2-mile drift dive can be made here. You sail along with the current just above the reef fingers. When you see something interesting, simply drop down behind the ledge projection and you will be out of the flow. After your inspection, simply push off into the current once again and you're off. There is plenty to see, including numerous pelagics; barracuda, African pompano and amberjacks are always around. Watch for large sea turtles; they love this area.

Delray wreck — Richard Collins

NO.15 DELRAY WRECK (Beach Dive)
Location: 150 yards off the south end of Delray's public beach.

This steel-hulled freighter, sunk in the 1920s, has become a very popular diving location that is excellent for the beginner. Many varieties of soft and hard corals have taken over the remains which are now in three distinct parts. The parts can be located by sighting the dark shadows on the white sand bottom. Depths to the sand are 22 feet. Macro-photography is excellent. This is a first-class beach dive.

Boat traffic is heavy, especially on the weekends. Be sure to fly your dive flag!

23
BOCA RATON
HILLSBORO INLET

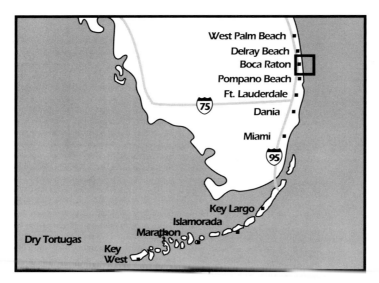

Offshore Diving Locations

1. Jap Rock
2. Boca North Beach Ledge
3. Westervelt's Ridges
4. Grouper Hole
5. Boca Artificial Reef Ledge
6. *Noula Express*
7. Separated Rocks
8. Berry Patch
9. Hillsboro Ledge
10. Labonte Reef

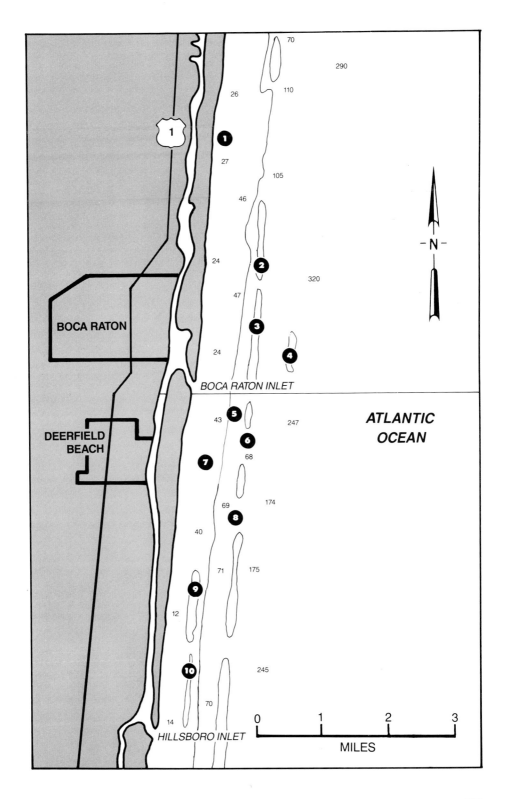

BOCA RATON

BOCA RATON INLET

DEERFIELD BEACH

ATLANTIC OCEAN

HILLSBORO INLET

-N-

0   1   2   3
MILES

NO.1 JAP ROCK (Beach Dive)
Location: North end of Boca Raton public beach.

This is a good reef for snorkelers. Its shallow rocks are full of tropicals and an occasional lobster. While it is excellent on calm days, on rough days this reef can be hazardous to your health! Waves breaking on the rocks (which make the site easy to locate) can cut the unprotected diver. This is also an unprotected section of beach. Watch for rip currents and check with local dive shops for conditions.

NO.2 BOCA NORTH BEACH LEDGE
Location: Seaward of Boca Radar Tower, about 1 mile north of the Boca South Beach Pavilion.

The North Beach Ledge is 10 to 15 feet long, in 60 feet of water. Sea whips, sea fans, sponges and hard corals top the ledge, while caves and overhangs provide food and cover for marine life. Moray eels grow to six feet here, and lobster abound during the winter months. Anchor on the edge, go north, and return to anchor. Fly your dive flag here!

Great Barracuda   Paul Humann

NO.3 WESTERVELT'S RIDGES
Location: Approximately 3/4 of a mile due east of Palmetto Park Beach Pavilion.

Westervelt's is a 15-foot-high ledge that drops on the outside to 70 feet. Overhangs and caves are common. Lobster, grouper and hogfish are abundant. This reef is noted for moray eels and many small caves.

NO.4 GROUPER HOLE
Location: Approximately 1/2 mile north of the Boca Inlet, about 1-1/2 miles offshore.

This deep reef lies in approximately 140 feet of water and, on an incoming tide, has visibility which can exceed 100 feet. Jewfish and grouper frequent this reef during the winter months. Drift diving will provide access to Grouper Hole. Be ready for the large critters! Lobster grow to around 10 pounds at this site.

NO.5 BOCA ARTIFICIAL REEF LEDGE
Location: Just south of Boca Inlet, approximately 3/4 of a mile offshore.

A popular artificial reef area adjacent to natural rock ledges. Depths to the sand are 70 feet. The landward side of the reef has a ledge approximately 10 to 15 feet high, running north and south. Visibility on an incoming tide can be from 50 to 100 feet. Fifty to 75 feet east of the ledge is a small artificial reef composed of erojacks stacked in piles. This is an excellent diving and fishing locale, which includes rockfish and midwater sport fish. Currents are often strong here!

NO.6 *NOULA EXPRESS*   14283.3   62085.4
Location: One mile offshore. ENE of the Deerfield Pier.

A 114-foot steel-hulled Danish freighter was sunk in 1988 as a joint venture of the Broward and Palm Beach County Artificial Reef Programs. She sits is in 71 feet of water.

NO.7 SEPARATED ROCKS
Location: South of Deerfield pier off the south end of the public beach.

From a depth of 40 to 45 feet in the sand, these large blocks of coral rise five to seven feet. These blocks provide excellent cover for lobster. Because this reef is on the southernmost tip of the artificial tire reef, you can drift for over two miles on an incoming tide. This reef can be reached only by boat, since it is too far a swim from the beach. Check with your local dive shop about conditions and trips to the reef.

NO.8 BERRY PATCH   14281.1   62088.5
Location: Almost one mile offshore inside the third reef.

A 65-foot steel tug was sunk in 71 feet of water on August 15,1987. A 40-foot boat and a 50-foot steel houseboat are located 110 feet to the southwest.

NO.9 HILLSBORO LEDGE
Location: 1/2 mile south of the last rock pile on Deerfield Beach.

A 5- to 10-foot coral ledge in depths of 34 to 40 feet, this is an excellent reef for the beginner. There are numerous tropicals to photograph or collect. Lobsters are scattered in rocky sections of the reef. Because of its location just outside the Inlet, there is not a long wait between suiting up and diving.

NO.10 LABONTE REEF
Location: Seaward of Hillsboro Landmark Condo, 1 mile north of Inlet.

The top of this reef is in 35 feet of water and reaches to a sandy bottom at 45 feet. This is a nice beginner's reef since it is relatively close to shore, easy to find, and contains an excellent variety of fish and invertebrate life. The inside edge is full of holes. Lobster, snapper and grouper can be found. Because this reef is too far to swim to from the beach, local dive shops run daily trips to Labonte.

24
POMPANO BEACH FORT LAUDERDALE

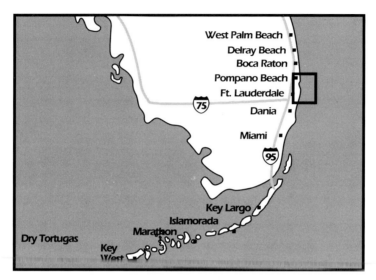

Offshore Diving Locations

1. Suzanne's Reef
2. Dorman Artificial Reef
3. Rodeo 25
4. Third Reef Ridge/Pompano
5. Pompano Drop-Off
6. Capt Dan
7. Lauderdale-by-the-Sea
8. Hogfish Ledge
9. Anglin Pier Ledge
10. Rebel
11. *Jay Scutti* Tug and Yachts
12. Robert Edmister Reef
13. Jim Atria Artificial Reef
14. *Mercedes*
15. Lauderdale Coral Gardens
16. Houseboat
17. Orange Reef
18. Charlie's Grotto
19. 82' Yacht *Monomy*
20. Osborne Reef
21. Yankee Clipper/ Erojack Reef
22. Spotfin Reef

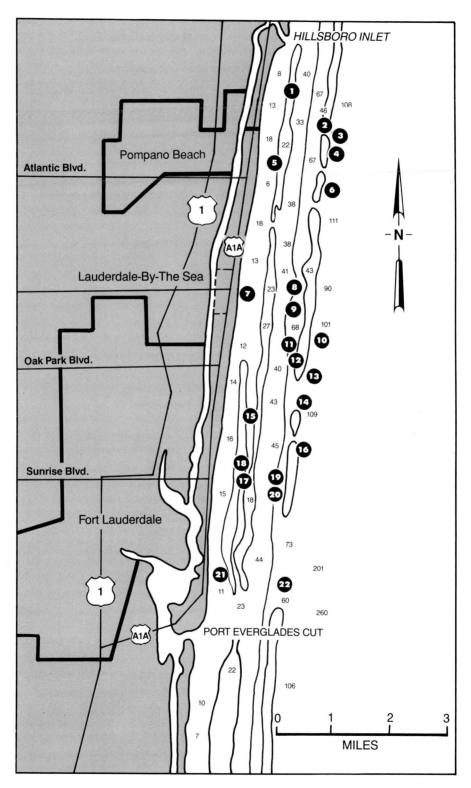

Area Information

The popular reef diving spots from Pompano Beach through the Ft. Lauderdale area are found on three distinct reef lines called the first, second and third reefs. These reefs generally run parallel to the shoreline.

The distance from the beach to the first reef varies from 100 to 300 yards. Depths range from 10 to 15 feet. The reef reaches a depth of 30 feet in a few spots. These patches of natural reef support a variety of marine life. Sea fans and sponges cover the top of the ridge. This is an excellent area for the novice scuba diver and is a delight for the free diver. Scuba diving is not allowed off the beach south of Oakland Park Boulevard, but snorkelers can make beach entrances anywhere along the beach. Few large game fish are found this close to shore, but multicolored tropicals are found by the thousands. The first reef provides excellent lobstering during the season. Many divers have grabbed their first bug from the pockets along the reef line. Visibility is not always the best, depending on the weather. Strong winds will kick up the sand in the shallow waters.

The second reef lies about half a mile offshore in most areas. Depths vary from 30 to 50 feet. Visibility is better than on the first reef and is nearly always good. Throughout most of the county the second reef is a limestone ridge with spurs and grooves running east and west across the reef. Soft corals and a variety of sponges decorate the limestone. You not only discover an abundance of tropicals, but large schools of grunts, spadefish, goatfish and yellowtails as well. The under water photographer will be able to find more than enough beautiful scenery for outstanding shots. The white sand that surrounds the reef is a prime hunting ground for shell collectors.

The third reef starts about a mile offshore and runs at depths from 60 to 100 feet. Because of the depths and sometimes strong currents, this reef should be explored by the more skilled diver. Larger game fish make this area their home and provide excellent spearfishing opportunities. Both the hard and soft corals are larger and more abundant in the deeper water. The visibility is nearly always outstanding. Drift diving is the most popular way to explore the third reef.

A highlight of Broward County diving is the opportunity to explore many exciting wrecks. There are over 50 artificial reef sites along the county's 24-mile coastline. This dramatic collection of vessels that attracts thousands of divers each year was acquired, cleaned and placed on the ocean floor under the direction of the Broward County Artificial Reef Program.

The summer months of May through September offer the best visibility and, of course, the warmer water temperatures. Several outstanding diving businesses are located in the South Florida area. They will be more than happy to provide interested divers with information about diving conditions and to arrange diving charters.

NO.1 SUZANNE'S REEF

Location: On the first reef line, about a quarter mile offshore and 3/4 of a mile north of the Pompano Pier, and the same distance south of the Hillsboro Inlet.

Depths vary from 10 to 18 feet on Suzanne's Reef, with the ledge from one to five feet in height. There is good shelling on the sand flats inside the ledge. Tropicals are numerous and lobstering can be good during certain times of the year. The inside

rock ledge is pocketed with small crevices, while soft corals and sponges grow on top of the reef. The outflow from the inlet will often lower visibility in the area.

NO.2 JAY DORMAN ARTIFICIAL REEF   14273.6   62096.1
Location: One mile directly offshore from the Pompano Pier (just north of Atlantic Blvd.)

The 130-foot luxury schooner *Panda* was sunk in May 1987 as an artificial reef memorial for Jay Dorman. The beautiful sailing vessel rests in 78 feet of water on her port side. It had formerly been a charter craft for Windjammer Cruises before fire damage in 1984. An 85-foot schooner *Alpha* and two Qualman tugs were put down about 50 yards north of the *Panda*. This area is called the Rodeo Divers Reef **14273.6 62096.3.**

NO.3 RODEO 25   14273.8   62095.3
Location: 1-1/2 miles due east of the Pompano Pier on the outside edge of the third reef.

A 215-foot twin-masted Dutch freighter was sunk in May 1990 to celebrate the Pompano Beach Fishing Rodeo's 25th anniversary. She lieson an east/west line in 122 feet of water. Her relief is 70 feet.

NO.4 THIRD REEF RIDGE/POMPANO   14272.0   62097.4
Location: 1-1/2 miles east-southeast of the Pompano Pier and straight out from the blue water tower.

Depths on top of the third reef average 50 feet and the visibility is usually good along the outside edge of the reef line where depths taper to 95 feet. The 12- to 15-foot-high outside wall is very scenic and therefore a popular area for underwater photography. Basket sponges line the top of the ridge, while multicolored sea fans, sea whips and other soft corals cover the outer edges. Grouper, barracuda, and chubs swim through the undercuts. This area can be subject to strong currents at times, so use caution.

Richard Collins

NO.5 POMPANO DROP-OFF

Location: About 1/2 mile due east of the large blue water tower on Pompano Beach.

This large, flat, rocky reef starts just 15 feet below the surface, and drops to just over 30 feet on the outside. Like most of the areas close to shore, visibility is usually poor, but there is little problem caused by currents. There are few large fish, but good tropical fish watching.

NO.6 *CAPT DAN*   14272.3   62096.9

Location: Just over a mile south from the Jay Dorman Artificial Reef. 1-1/2 miles from shore; just outside the third reef line.

The 175-foot Coast Guard buoy tender *Hollyhock* was sunk in February 1990 in memory of Capt Dan Garnsey. She lies in 110 feet of water in the Rodeo Reef Site.

NO.7 LAUDERDALE-BY-THE-SEA (Beach Dive)

Location: Turn east off I-95 onto Commercial Blvd. Go straight 3.3 miles. After crossing the intersection with A1A, go one block and turn left on El Mar. In one block, turn right into parking area after stop sign.

The one-mile stretch of beach is considered one of the state's best beach diving sites. At the end of Commercial Blvd., the Anglin Pier extends out to sea. The best diving is 100 yards north or south of the pier. Remember that you cannot dive near any pier in Florida waters.

The first line of hard bottom rises only one to two feet off the sand. It is sparsely covered with soft corals and sponge. It is located halfway down the pier's length in eight to 10 feet of water. You can line up with the pier house.

The end of the pier marks the first true reef line. Depths here are to 30 feet on the eastern edge. Large angelfish, small Nassau grouper and snapper are all along this reef section. The reef rises off the sea floor four to eight feet and is adorned with sea fans, sponge and soft corals.

One hundred yards out from the pier is the third reef. The best spots here are described in the next two reef listings: Hogfish Ledge and Anglin Pier Ledge.

Beach diving is best when the wind comes from the northwest. The surf can at times be calm as a lake. With less surge the visibility improves markedly. Be sure to check tide tables so that you will not have to return to the beach during an outgoing tide. Always trail a float (such as an inflated tire tube) with a LARGE dive flag to mark your underwater location. Boat traffic is heavy along all sections of this reef. Always listen, look and ascend with caution. Many boaters, unfortunately, fail to heed the diver's flag. The Marine Patrol constantly patrols the area and will check for illegal lobster or divers not tethered to a visible dive flag.

This is also a good place to night dive from the beach. You can mark your position by the lights on the pier or from hotels along the beach. It would be a good idea to attach a warning strobe light to your surface float.

NO.8 HOGFISH LEDGE   14269.8   62100.3

Location: 6 miles north of Port Everglades Cut. Just north and a mile and a quarter out from Anglin Pier at Lauderdale-by-the-Sea.

This is a nice patch reef with depths to 60 feet. Plenty of fish make their home here along with sea turtles, morays and lobster.

NO.9 ANGLIN PIER LEDGE

Location: Just south of Hogfish Ledge on the third reef line.

This is another patch reef section with depths to 60 feet. A good spot for drift diving; currents are usually strong.

NO.10 *REBEL*   14267.1   62103.0

Location: Five miles from the Port Everglades Cut, just outside the third reef line.

During the summer of 1985, a 150-foot Norwegian freighter, built in 1947, was placed on the sea floor to become another spectacular artificial reef site. Her original name *Andrea* was changed to the *Rebel*. She sits even keel in 110 feet of water, with her bow pointing directly north. It is 80 feet to her deck. Buoyed descent lines are sometimes in place. This is an open wreck easily penetrated.

NO.11 *JAY SCUTTI* TUG AND YACHTS   14265.2   62106.3

Location: 4-1/2 miles north of the Port Everglades Cut.

This is an excellent wreck site in only 67 feet of water. The 100-foot tugboat sits between two sleek sailing hulls. The 95-foot sailboat *Pride*, sunk in 1987, is 100 feet south of the tug **14265.2   62106.3**. A yacht, *B.H. Lake*, is 150 feet northeast.

NO.12 ROBERT EDMISTER REEF   14273.4   62107.0

Location: 300 yards due south of the Jay Scutti.

The 95-foot Coast Guard Cutter *Cape Gull* was sunk on December 1989 in 70 feet of water.

NO.13 JIM ATRIA ARTIFICIAL REEF   14266.5   62103.5

Location: 4 miles north of the Port Everglades Cut; just outside the third reef line.

In September 1987, the Broward County Artificial Reef Program sank the 240-foot freighter *Poinciana* in 112 feet of water. She lies on her port side. Depths to the hull are 75 feet.

NO.14 *MERCEDES*   14265.2   62105.2

Location: Approximately 4 miles north of the Port Everglades Cut, in line with the outside edge of the third reef.

The *Mercedes*, a 198-foot freighter, was beached during a storm on November 23, 1984. She came to rest near the oceanside pool of a Palm Beach socialite. The event was well-recorded by the national television networks. It took over three months and a quarter million dollars to remove the vessel.

The derelict freighter was purchased by the Broward County Environmental Quality Control Board who, with the help of the South Florida SCUBA Divers Club, prepared the ship for sinking. She was placed down on March 30, 1985, off Fort Lauderdale Beach, a half mile north of Sunrise Blvd.

She now rests even keel in 90 feet of water. The deck is at 60 feet and her tower starts at 45 feet. Just to her east the bottom drops to 100 feet; to the west 95 feet. Several exit holes were cut in her hull and bulkheads to enable fish to enter and exit the vessel. Plenty of sunlight brightens the interior. Often there is a current that runs north, at 1 to 1-1/2 knots that must be negotiated by divers. It is generally stronger near the surface. This condition, plus the depth, combine to make the *Mercedes* an advanced open-water dive.

Most divers drift the wreck with the current. Three descent lines are sometimes in place. They run from buoys to the bow, tower and stern. The wreck is filled with millions of tiny baitfish which attract barracuda, jacks and other open-water feeders to the site. This is an extremely popular dive location, so expect a crowd on the weekend.

The *Mercedes* is part of a triangle of wrecks that are within a half mile of each other. The *Mercedes* is to the south, the *Jay Scutti* to the west, the *Rebel* to the north and the *Jim Atria* in the center.

NO.15 LAUDERDALE CORAL GARDENS
Location: 3-1/2 miles north of the Port Everglades Cut, out from the north end of Birch State Recreational Area. The patches are just inside the second reef line.

This is a beautiful patch reef area, well over 100 yards in diameter. It is excellent for shallow scuba or snorkeling with depths from 12 to 30 feet. Colonies of hard corals are developing on the limestone substrate; a rarity this far north. Several small stands of rare pillar coral are growing on the site. The reefs are in a direct line with an active turtle run. Night diving is easy and exciting. Schools of squid and octopus are often seen after dark.

NO.16 HOUSEBOAT   14263.7   62107.0
Location: Just south of the Mercedes.

The 50-foot houseboat rests in 83 feet of water in a patch reef area. It has an 18-foot profile. Just east, depths go to 100 feet. Spearfishermen and lobster hunters favor this spot because of the large number of sea creatures that roam here. A triangle of three broken-up boats is located 200 yards northeast.

NO.17 ORANGE REEF
Location: 3 miles north of the Port Everglades Cut, directly offshore from the orange roof of the Howard Johnson Motel.

After diving the deep wrecks this spot provides a nice shallow dive with depths from 20 to 30 feet. The reef's main feature is a coral ledge that is 10 feet high and over 50 feet long. Several other reef fingers run east toward the sea. Their valleys are filled with interesting sea life.

NO.18 CHARLIE'S GROTTO
Location: 125 yards north of Orange Reef.

A beautiful sand grotto bordered by a 10-foot reef wall is located on the second reef line. Depths are 30 feet in the sand. There is seldom any current in this area. Nice hard corals, including elkhorn and staghorn, grow in the ledge. Barrel sponges are common.

NO.19 82' YACHT *MONOMY*   14263.2   62107.5
Location: About 1-1/4 miles due east of Sunrise Boulevard. 1000 yards north of Osborne Reef.

An old wooden yacht lies on its side in 55 feet of water on the outer edge of the third reef. Most of her portholes and brass fittings have been removed, but it is still fascinating to explore her broken remains. The yacht has become an artificial reef, and is in an advanced stage of development. Snapper, barracuda, jacks and other mid-water sport fish are commonly sighted. The depth of the surrounding reef ranges from 45 to 70 feet.

NO.20 OSBORNE REEF   14263.3   62107.9

Location: 1-1/4 miles east of Sunrise Boulevard.

This large artificial reef is made up of the wreckage of old barges, tires and concrete erojacks. Osborne is an excellently developed artificial reef established in the 1970s. It supports an almost unbelievable abundance of fish life. Depths vary from 70 feet in the sand to 60 feet on top of the reef.

NO.21 YANKEE CLIPPER/EROJACK REEF   14259.2   62115.5

Location: Due east of the Yankee Clipper Hotel.

Huge piles of concrete jacks provide a home for a multitude of sea life in 15 to 25 feet of water. An advanced life cycle has developed within the rubble, creating a most interesting dive.

NO.22 SPOTFIN REEF   14258.9   62112.6

Location: One mile from the beach on the third reef, 1-1/4 miles northeast of Port Everglades Inlet.

This is a very pronounced ledge starting in 50 feet of water and dropping to 65 feet in the sand. The ledge is picturesque, undercut in many places with caves and crevices. The area is covered with a large variety of soft and hard corals and is good for spearfishing and tropical fish collecting.

25

DANIA – HOLLYWOOD

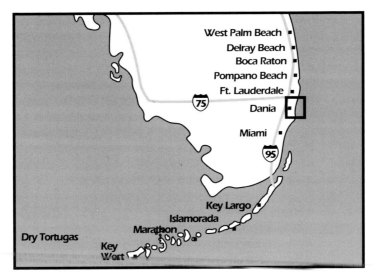

Offshore Diving Locations

1. John Lloyd Reefs
2. Barracuda Reef
3. Hammerhead Reef
4. Dania Erojacks
5. Dania Beach Reefs
6. Glenn's Reef
7. Hollywood Beach Reefs
8. Hojo's Reef
9. Johnson Street Reef
10. Tenneco Towers

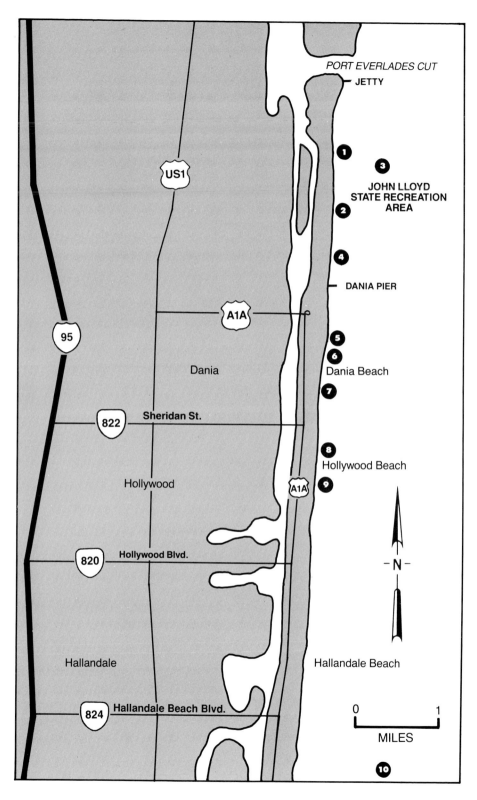

PORT EVERLADES CUT

JETTY

US1

JOHN LLOYD
STATE RECREATION
AREA

DANIA PIER

A1A

95

Dania

Dania Beach

822 Sheridan St.

Hollywood

A1A

Hollywood Beach

820 Hollywood Blvd.

– N –

Hallandale

Hallandale Beach

824 Hallandale Beach Blvd.

0   1

MILES

Area Information

The largest concentration of good beach diving in Florida is found off the three beaches that start south of the Port Everglades Cut.

John U. Lloyd State Recreational Area—just over two miles of beachfront—extends from the jetties on the south bank of the cut to the Dania pier. Lifeguards patrol the entire strand. There is a fee of $1.00 per vehicle and $.50 per passenger to enter the 244-acre barrier island park. Access is from Dania Beach Blvd. (A1A).

Dania Beach—1,000 feet of beach. Access is east off A1A on Oak Street.

Hollywood Beach—five miles of beach stretch south from Sherman Street to Greenbrier Street. a developed area with five access walkways across the dunes. Lifeguards are on duty.

The next beach south is Hallandale. Although the reef lines continue to parallel the beach here, they run further offshore.

The three ledge reef lines that parallel the coast south from West Palm Beach continue to Miami. From Dania to south Hollywood, the first ledge is located close enough to the coast to make beach diving practical. Both scuba and snorkeling are legal. Spearfishing is allowed outside the bathing area, but the shaft must be carried in one hand and the gun in the other when swimming out or coming in. Lobstering is a popular activity during the season.

Displaying the dive flag is not only the law but also common sense. There is a heavy concentration of power boats and windsurfers along all these beaches. Most divers tether themselves to a float (such as an inflated tire tube). The diver's down flag should fly at least three feet above the float.

The first limestone ridge rises one to two feet off the bottom in most places, but often disappears under the sand for brief stretches. In most spots, it is found 75 to 100 yards off the beach. The first true reef is more pronounced, rising four to five feet in places. Both lines support an extensive marine life community. Expect to see thousands of tropical fish, crustaceans, sponges, sea rods, whips and fans. Pockets of hard corals become easier to find as you go south. Depths on the first ledge are 10 to 15 feet and to 30 feet on the reef. Visibility depends on the weather. The summer months, when the wind is slack, are the best time to beach dive. Lifeguards working the beaches are your best source for diving information. They know the tides, currents and hazardous areas to avoid, and can point out the best entry points for all the reefs.

NO.1 JOHN LLOYD REEFS (Beach Dive)

Location: From I-95 take Sheridan Street east (look for the water tower with Hollywood lettered on the side). In 3 miles you cross the Intracoastal Bridge. Turn left on A1A and drive to the A1A loop. The park entrance is just north of the loop. Park in the first parking area inside the park and cross the wooden ramp to the beach. Swim straight off the beach until you are even with the end of the Dania pier, just to the south.

This is a nice section of the first reef system that rises four feet off the bottom in 20 feet of water. It is topped with sponges and sea fans. The ledge runs north and south. The northern section breaks down into a rocky area good for lobster hunting.

Good beach diving can also be found in the park's northern section, 200 yards south of the jetties. There are two parking lots convenient to this area. Spearfishing is not permitted near the jetties.

NO.2 BARRACUDA REEF   14255.9   62117.8

Location: One half mile out from the John Lloyd State Recreational Area. Marked by ten mooring buoys. The northern buoy is south of the Port Everglades Cut and the southern buoy is just north of the Dania Pier **14255.4 62118.1.** *The buoys are spaced 160 feet apart.*

Barracuda is a high-profile reef ledge that rises 10 to 15 feet off the bottom. The depth is 20 feet to the top, 35 feet in the sand. The limestone ledge is topped with sea fans and sponges and has numerous undercuts. Fish life is plentiful. This is an excellent spot for fishwatching. The mooring buoys on Barracuda are the first to be installed in Broward County. The project was sponsored by the Ocean Watch Foundation and is maintained by the Florida Ocean Sciences Institute.

NO.3 HAMMERHEAD REEF

Location: One half mile south of the Port Everglades Cut on the third reef line.

This is an excellent close-in boat dive, starting at Snook Alley and running 2-1/2 miles to the Dania Pier. The reef rises over 18 feet. Depths to the sand are 60 feet shoreside, and nearly 90 feet on the oceanside. This is a great place for a drift dive. You are guaranteed to see a variety of interesting sea life.

NO.4 DANIA EROJACKS (Beach Dive)   14254.2   62123.0

Location: In the John Lloyd State Recreational Area, go 1/2 mile to the first parking lot (this is easy to miss). Park at the south end and walk across the first wooden ramp. Follow the beach south, past the restroom facility and forest of tall Australian pines, to a 4-foot-tall Milan palm standing alone (it is the only one). Swim straight off the beach from the palm. The erojacks begin 3/4 the distance to the pier's end.

The rubble pile runs parallel with the pier in 10 to 20 feet of water. The large, coral-encrusted concrete reef sections have been down for many years and support a variety of marine life. Erojacks are cement beach barriers that were used during WWII. They are four to five feet in diameter and resemble children's jacks. They were place down years ago to help slow beach erosion. At the far end is a nice shelf reef with ledges. Gorgonians and sponges are prolific. Larger angelfish and schooling spadefish constantly hang around this spot. There are a few lobster and many moray eels. It will take an average diver a single tank of air to make the long swim around the tip and back to shore. Be sure that your dive flag is displayed prominently; there is heavy boat traffic in the area.

NO.5 DANIA BEACH REEFS (Beach Dive)

Location: From I-95, take Sheridan Street east for 3 miles to A1A. Go north. Take the first right before the Intracoastal Bridge overpass (just south from the entrance to the John Lloyd State Recreational Area). You will see the Sea Fair, a tall pink and blue building. Park in the southern section of the parking area. This is the only access for diving off Dania Beach.

Dania Erojacks

Richard Collins

Swim east for a pier length and a half where a rubble reef area begins. Continue out until you reach a 5-foot ledge in 20 to 25 feet of water. The reef runs to the pier, but stay away because of the danger from monofilament line entanglement. The far side of the reef is another 100 to 200 feet east. This site has a lot of sea life.

NO.6 GLENN'S REEF (Beach Dive)
Location: Near the parking area described above, there is a white motel with a conic tower. Take a sighting off the tower and swim directly off the beach until lined up with the pier's end.

This is a beautiful reef section that is excellent for snorkeling and photography. The site becomes milky when the current runs north on an ebb tide.

NO.7 HOLLYWOOD BEACH REEFS (Beach Dive)
Location: From I-95, go east on Sheridan Street for 3 miles to A1A. Go north for 1/2 mile to Forrest Street. Turn north on N. Surf Road (beach service road). Metered parking is available from Forrest Street north to Dania Beach.

From North Beach north through Dania Beach, the first reef line runs almost continuously. Two popular beach diving spots are off Green and Charleston Streets. The reef is low profile, but filled with ledges and crevices. Depths are from 10 to 20 feet. Gorgonian are the primary coral. These sites are good for collecting tropicals and macro-photography. There are plenty of angelfish, spadefish, yellowtails, tangs and some lobster. Visibility varies depending on currents and wind conditions.

NO.8 HOJO'S REEF (Beach Dive)

Location: From I-95 go east on Sheridan Street for 3 miles to A1A. Turn south and go 1/2 mile to Howard Johnson's (located between Carolina and Taft Streets). Limited metered parking is available.

The first reef line runs less than 75 yards from the beach here. It is a broken low-profile reef covered with tropicals. Spanish and shovelnose lobsters can sometimes be found in the area.

NO.9 JOHNSON STREET REEF (Beach Dive)

Location: From I-95, go east on Sheridan Street for 3 miles on A1A. Go one mile south to Johnson Street.

This is the last reef section close enough to shore to be considered a beach dive. The reef line is over 110 yards off the beach.

NO.10 TENNECO TOWERS   14246.9   62122.7

Location: Nearly a mile off Hallandale Beach Blvd., just north of the Dade County line. All the sections line up with the Hallandale water tower directly in front of the radio antenna.

The Tenneco Oil Company, at considerable effort and expense to themselves, transported and sank two huge Gulf of Mexico oil production platforms and a section of a third on the sea floor just out from the Dade and Broward county line. The rigs were placed down in five sections on October 3, 1985, creating the largest artificial reef in southeast Florida. Three decked sections are within the sport diver's limits. Two are in 110 feet of water and rise to within 60 feet of the surface. The lower deck is 20 feet below the upper deck. A third, somewhat smaller platform, rests at 97 feet, topping out at 65 feet. It is the closest to shore. The second is about 100 yards seaward from the first, and the third is out another 85 yards. The two deep water sections which consist of supporting legs, called jackers, rise 80 feet from a depth of 190 feet **14247.3   62120.9.**

When visibility and currents are right, the Tenneco Towers are one of the most spectacular dives in Florida. The flow of nurturing currents that move freely through the structures have covered the rails, stairways and underside of the deck with a jungle of soft corals. Even hard corals have begun to grow in profusion on the grating. Fish life swarms around the great structures. Colorful Spanish hogfish, queen angels and spotfin butterflyfish dine on tiny invertebrates hidden in the growth, while large pelagic jacks cut through the towers in pursuit of baitfish. What a dive! Thanks, Tenneco.

26

MIAMI

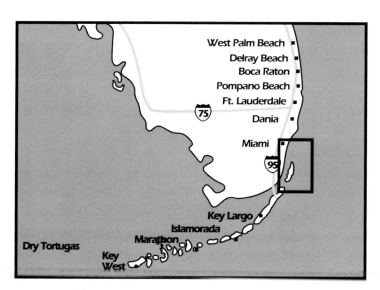

Offshore Diving Locations

1. Haulover Shallow Rocks
2. Harbor House Reef
3. *Narwal*
4. *Andro*
5. Crane Wreck
6. Eighty-Eighth Street Reef
7. *Deep Freeze*
8. Koppin Memorial Reef
9. Ben's Antenna Reef
10. Billy's Barge
11. *Shamrock*
12. Schoolmaster Ledge
13. Key Biscayne
14. Schooner Wreck
15. Emerald Reef
16. Biscayne Wreck/Banana Freighter
17. *Miracle Express*
18. *Sheri-Lynn*
19. *South Seas*
20. Cluster of Small Vessels
21. *Rio Miami*
22. *Belzone One*
23. *Arida*
24. *Proteus*
25. *Lakeland*
26. *Orion*
27. Belcher Barge #27
28. *Ultra Freeze*
29. *Tarpoon*/Kevorkian Memorial Reef
30. Hopper Barge
31. *Steane D'Auray*
32. Fowey Rocks
33. Sonar Structure
34. Brewster Reef
35. *Blue Fire*
36. Star Reef
37. Mystery Reef
38. Triumph Reef
39. *Almirante*
40. Long Reef
41. Ajax Reef
42. Doc DeMilly Memorial Reef

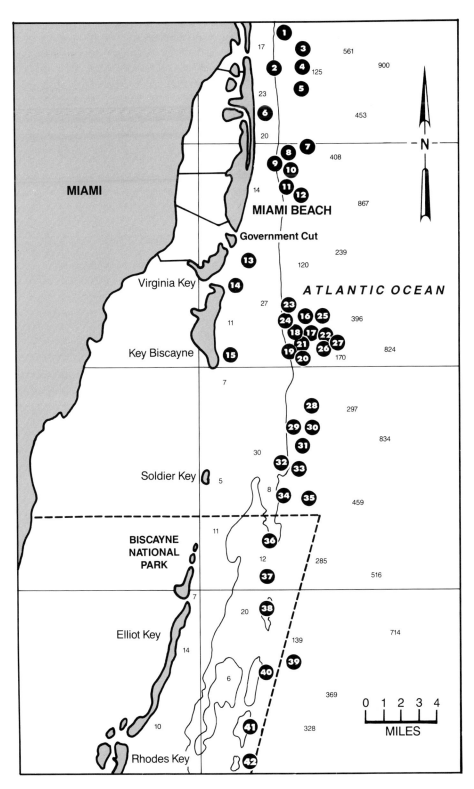

MIAMI

MIAMI BEACH

Government Cut

Virginia Key

Key Biscayne

Soldier Key

BISCAYNE
NATIONAL
PARK

Elliot Key

Rhodes Key

ATLANTIC OCEAN

N

0 1 2 3 4
MILES

Area Information

Miami's offshore waters have long been a delight to water sportsmen. For years the city has been the center of many water-related activities. Sport diving, of course, is no exception.

Recent artificial reef construction projects by the Dade County Environmental Resources Management office (DERM), Dade Sportfishing Council and private concerns have produced some of the state's most dramatic diving sites. Large derelict ships have been purchased, cleaned and towed to specially permitted sites where they were sunk. The large structures quickly attract myriad sea life, bountiful for fishermen and thrilling for divers to explore. Because of the clear waters that bathe Miami's wrecks, their visual impact is startling. Water visibility varies, depending on the caprice of the Gulf Stream. Days with 100 feet of visibility are not uncommon. Windy periods seem to have only marginal effect, but water run-off from Miami's street system after heavy rains can drop visibility considerably.

Depths vary from a few feet at shallow-water reef areas to several hundred feet only a few miles offshore. The wrecks are all relatively deep—from 70 to 200 feet. Those wishing to visit these sites should be experienced open-water divers. Although spearfishing is now allowed on wreck sites, county ordinances banning such activities will probably be implemented in the near future.

Natural rocks and coral reefs are prevalent from North Miami south through Biscayne National Park. The southern reefs are as lovely as any found in the Caribbean. Acres of antler corals, teeming with schooling fish, lie in wait for visitors.

Some of the nation's oldest and best operated diving stores and charter boats operate in the Miami area. They know the surrounding waters well and are always ready to provide sport divers with assistance in discovering Miami's spectacular underwater world.

To obtain a small waterproof booklet with visual ranges and Loran-C numbers for Miami's artificial reef sites, send $19.95 to: Artificial Reef Program, 111 N.W. 1st St., Miami, Florida 33128. A video entitled Dive Miami—The Best of Dade County's Artificial Reef Program is also available from the same address for $19.95.

NO.1 HAULOVER SHALLOW ROCKS
Location: Directly east of 158th Street and the beach. If you run a north and south course, you will pass over a covered discharge pipe with an elevation of 10 to 15 feet.

The area has scattered coral formations on a white sand bottom that varies in depth from 30 to 40 feet. Grouper and small snapper are common. The rocks that cover the discharge pipe are a protective haven for an array of colorful tropicals. The best diving in the area is at high tide.

NO.2 HARBOR HOUSE REEF
Location: Straight out from the Harbor House (first large apartment complex south of the Haulover Inlet on Miami Beach).

The reef starts in 45 feet of water and rapidly drops to 85 feet. The reef system supports numerous varieties of soft corals and sponges. This is an excellent area to photograph grouper, turtles and large green morays.

NO.3 *NARWAL*   14239.7   62127.8

Location: Just under 2 miles east of the Haulover Inlet in the center of the Haulover Artificial Reef site.

On April 17, 1986, the 137-foot steel freighter Narwal was sunk in 115 feet of water. Her profile is 28 feet. She is sitting upright and intact, including rudder and prop. Holes were cut in the bulkheads, opening the hull compartments to divers. This area is not often visited by charter dive boats so this is a chance to have a wreck to yourself.

NO.4 *ANDRO*   14237.7   62129.4

Location: 2 miles east and slightly south of the entrance to the Haulover Inlet near the center of the Haulover Reef Site.

The *Andro* is a classic wreck! The 165-foot vessel was built in 1910 as a luxury yacht. Originally outfitted with electric motors she was converted to diesel during WWII when she became a subchaser. After a customs seizure, the steel freighter was purchased by DERM and put down in 103 feet of water on December 17, 1985. Her 38-foot profile has classic lines—a high bow, rounded-off flare tail, and twin propellers.

NO.5 CRANE WRECK   14237.9   62129.8

Location: 2 miles east of the 7th building south of Haulover Cut.

The wreckage of an old steel crane rests on a sand bottom in 80 feet of water between coral fingers. Visibility is usually between 30 and 40 feet. Spearfishing is common in the area due to the depth and expansive reef formations. Diving is exciting and challenging. The area is not recommended for inexperienced divers.

NO.6 EIGHTY-EIGHTH STREET REEF

Location: The reef lies about 3/4 of a mile due east of the Holiday Inn on 88th Street and Collins Avenue.

Coral formations are found at a depth of 35 to 40 feet. The site abounds with tropical fish and is a favorite with fish collectors. On the deeper side of the reef, larger fish and lobster can be found among the bigger coral heads. Always fly a diver's flag, since this is a heavy boat traffic area.

NO.7 *DEEP FREEZE*   14230.8   62134.4

Location: About 5 miles north-northwest of the Government Cut jetties in the Pflueger Artificial Reef Site.

A 210-foot freighter is resting east to west on her keel in 135 feet of water. Her top deck is at 100 feet. She has been down since October 1976 and is covered with algae and corals. All the companionways are open, but be careful of the fine silt on the deck. Be sure to carry a sharp knife on this and all wreck dives in case of fishing line entanglement. This ship is host to many large fish, including grouper, jacks, and barracuda. Moray eels and a large colony of Atlantic spiny oysters can be seen.

NO.8 KOPPIN MEMORIAL REEF   14229.5   62136.6

Location: 100 yards northwest of Billy's Barge in the Anchorage Site.

This site was started with the placement of a 75-foot steel barge as a memorial to a Miami motorcycle patrolman. Next to the barge are 32 8-foot concrete pipes and six 90-foot concrete girders from the old Rickenbacker Causeway **14229.3 62137.5**. The depth is 45 feet.

Due north is the *African Queen*, a 57-foot concrete boat sitting in 45 feet of water. **14229.3 62137.7.**

NO.9 BEN'S ANTENNA REEF   14229.3   62137.1

Location: Due east of Billy's Barge in the Anchorage Site.

Ben's Antenna Reef.   Richard Collins

In 50 feet of water, 15 pyramids have been formed using sections of radio towers. Each rises 20 feet off the bottom. The pyramids were designed by Ben Mostkoff, the coordinator of DERM's Artificial Reef Program, as an experiment to attract sea life.

North of Ben's Reef is the 100-foot steel barge, *Leon*. The barge has a 12-foot profile **14229.3 62136.9**. Just east of the barge is a large cluster of materials. In the center is the 70-foot steel Haitian cargo boat, *Esjoo*, also in 50 feet of water **14229.3 62136.7**. Next to it is a 55-foot wooden Haitian boat and 400 tons of concrete pipe, each 8 feet in diameter **14229.3 62137.1**.

Recent sinkings in the Anchorage Reef site are: 300 tons of 8-foot diameter concrete pipes **14229.3 62137.5**; concrete and tanks in 45 feet of water with a 16-foot relief **14228.9 62137.8**; the *Miss Karline*, an 85-foot steel vessel with a 15-foot profile **14229.5 62136.6**; the 65-foot steel tug *Patricia*, sunk in the summer of 1990, in 53 feet of water **14229.5 62136.6**.

NO.10 BILLY'S BARGE   14228.9   62137.2
Location: Approximately 200 yards north of the Shamrock in the Anchorage Site.

A 110-foot steel barge rests in 50 feet of water surrounded by 360 tons of concrete pipe. Concrete materials have been found to be excellent reef builders. Both hard and soft corals quickly attach themselves to the cement surfaces.

NO.11 *SHAMROCK*   14229.0   62137.4
Location: Approximately 3-1/2 miles north of Government Cut near the southern boundary of the Anchorage Artificial Reef Site.

The *Shamrock* is a 120-foot steel troop landing craft that was put down on June 28, 1985. The craft has a 16-foot profile off a 45-foot bottom. Water visibility around this vessel and other dives in the Anchorage Site is usually 30 to 40 feet.

NO.12 SCHOOLMASTER LEDGE   14227.6   62138.0
Location: In the southern section of the Anchorage Artificial Reef Site, just south of the Shamrock.

A 6-foot high ledge runs north and south for a quarter mile. It is 35 feet on top and 42 feet in the sand. Clouds of fish run the reef including large schools of porkfish.

NO.13 KEY BISCAYNE
Location: Just east of Key Biscayne, starting about a mile south of Government Cut.

This is a large area of patch reefs and coral mounds scattered over a sand bottom varying in depth from 15 to 40 feet. The clarity of the water usually allows you to easily spot the darker coral heads contrasted against the white sand bottom.

Because of the generally calm waters and varieties of depths in this locale, both experienced and novice divers will find this area suited to their talents. Sea fans and gorgonians are plentiful. Tropicals as well as game fish are common.

NO.14 THE SCHOONER WRECK
Location: 75 yards northwest of the red marker #2 on the south side of the Bear Cut channel near Key Biscayne.

This is a great place for the inexperienced diver to visit a wreck. The wreck lies in 10 feet of water on a smooth, grassy bottom.

NO.15 EMERALD REEF
Location: On the Atlantic side of Key Biscayne, toward the south end.

This is by far the most beautiful shallow reef group in the Miami area. The patch reefs are in 15 to 20 feet of water. Many of the better spots are difficult to locate but well worth the effort. Colors here in the shallow, clear water are truly breathtaking. Sponges in every color predominate. They are complemented by delightful patches of living coral. Every coral head and sponge teem with juvenile tropicals in the spring and summer. There are also many larger fish, including parrotfish, grouper and hog snapper.

NO.16 BISCAYNE WRECK/BANANA FREIGHTER   14218.5   62145.0
Location: The wreck is 4-1/2 miles east of Key Biscayne in the midwestern sector of the Department of Environmental Resources Management (DERM) Key Biscayne Artificial Reef Site.

A 120-foot freighter, once used for hauling bananas from Central America, sat derelict in the Miami River until 1974 when she was sunk to form an artificial reef. She rests in 55 feet of water and has a 15-foot relief. This is a very popular wreck dive because of the rather shallow depths and the breathtaking concentration of fish life constantly moving about her structure. At times, the baitfish schools are so numerous and immense that they shroud her from view. This is a fine place for a "fantasy" night dive.

NO.17 *MIRACLE EXPRESS*   14218.5   62145.0
Location: Touching the southern edge (stern) of the Biscayne Wreck.

A 96-foot interisland freighter was sunk without explosives in July of 1987 just south of the Biscayne Wreck. The power of the north-running currents have pushed the vessel next to the Biscayne. Her superstructure is intact, rising 25 feet off the sand bottom.

NO.18 *SHERI-LYNN*   14218.5   62144.5
Location: In the Key Biscayne Artificial Reef Site, 300 yards due east of the Biscayne Wreck.

Don't miss this wreck: it's incredible! The 235-foot steel freighter was placed in 95 feet of water. She is completely intact with a 45-foot profile to the top of the pilot house. This is a great wreck for photography due to the ship's stature. The structure is completely open with 3x4-foot holes cut through several bulkheads. The cargo area is immense.

Due north of the *Sheri-Lynn's* stern is a cluster of smaller vessels, including a 75-foot wooden shrimper, a 40-foot houseboat and a 100-foot barge.

Just south of the *Sheri-Lynn's* port bow is the first of 50 Chevron tanks. Each cylinder is 30 feet long and 8 feet in diameter, with the ends cut out. Included in this 100-yard diameter field of tanks are 20 cement mixer tanks, each 10 feet in diameter. Carefully plan your dive before exploring this large area. It is easy to be so captivated by all there is to see that you forget your bottom time. Currents can also be strong in this area.

Biscayne wreck.   Richard Collins

NO.19 *SOUTH SEAS*   14218.2   62144.7

Location: The wreck is 4-1/2 miles east of Key Biscayne in the southwestern sector of the Key Biscayne Artificial Reef Site, about 1/4 mile south of the Biscayne Wreck.

The *South Seas* was a sister ship to Hitler's personal yacht. During the early 60s she was used as a piano bar before sinking at the dock where she was moored. The 175-foot steel-hulled vessel now rests on a north/south line in 70 feet of water. She has been down since 1983. Today, her 15-foot profile is an exciting fish haven that, unfortunately, is not often dived

NO.20 CLUSTER OF SMALL VESSELS   14218.6   62144.4

Location: Found in the Key Biscayne Site, about 150 yards east of Biscayne Wreck.

This is a cluster of seven vessels in 85 to 100 feet of water: *Sarah Jane*, a 65-foot wooden boat; a 100-foot steel barge; a 60-foot wooden drift boat; a 40-foot steel houseboat; a 35-foot fiberglass boat; a 40-foot wooden vessel; and a 25-foot steel hull. The maximum relief is about 12 feet. Baitfish, jacks and grouper abound among the broken wreckage.

NO.21 *RIO MIAMI*   14218.4   62144.6

Location: In the Key Biscayne Artificial Reef Site between the Miracle Express and the Sheri Lynn, 100 yards from each.

This 105-foot steel tug was sunk in November 1989 in 67 feet of water. She is upright, listing slightly to starboard. Her cabin, with ladders and companionways intact, rises 30 feet off the bottom. Her sinking was featured on the ABC television show 20/20 hosted by Hugh Downs. In fact, Mr. Downs set off the detonator that sent her to the bottom, and was also the first to dive on the wreck.

NO.22 *BELZONE ONE*   14217.8   62144.9

Location: In the Key Biscayne Artificial Reef Site east of the Belcher Barge.

This 85-foot steel oceangoing tug was sunk in May 1990 in 68 feet of water. She is sitting upright with her search and navigational light still intact.

NO.23 *ARIDA*   14219.7   62143.7

Location: In the northwest corner of the Key Biscayne Artificial Reef Site.

The 165-foot *Arida* was an old LCT later converted into a freighter. Her steel hull is now resting on its side in 88 feet of water, with a 25-foot relief. Grouper and other large fish inhabit her remains.

NO.24 *PROTEUS*   14218.7   62144.3

Location: In the Key Biscayne Artificial Reef Site, 3/4 of a mile south of the Arida, or 800 yards northeast of the Biscayne Wreck.

The *Proteus* is spectacular due to its sheer size and huge propeller. It is a 220-foot steel freighter with its superstructure removed so that it could be placed in 72 feet of water. Her deck is 50 feet below the surface. The wreck was put down on July 24, 1985, with help in the clear-up provided by the diving certification organization NAUI.

NO.25 *LAKELAND*   14218.6   62143.8

Location: On the outside of the Key Biscayne Artificial Reef Site.

The 200-foot steel freighter lies east to west in 126 to 140 feet of water. She is keel-up, resting on her superstructure with a 30-foot relief. Her decks are filled with baitfish, barracuda, grouper and, occasionally, a shark.

NO.26 *ORION*   14217.4   62145.0

Location: The southwest corner of the Key Biscayne Artificial Reef Site.

The 118-foot steel tug once operated in the Panama Canal. She now rests in 95 feet of water with her twin sets of engines and ten-foot propellers intact. Since she sank in 1981, she has been one of Miami's most popular wreck dives. With stacks rising 30 feet and superstructure in place, she casts a haunting spirit over all visitors.

NO.27 *BELCHER* BARGE #27   13217.8   62145.1

Location: In the Biscayne Artificial Reef Site, 200 yards west of the Orion.

A 195-foot steel barge, filled with 500 tons of prefab concrete, was sunk on November 29, 1985, when four explosions opened her corners. The barge flipped on her way to the bottom. She now sits upside-down in 58 feet of water, on her concrete load. Experienced divers with lights can enter the corner holes and swim the barge's length, passing through the twelve 3x4-foot holes that were cut in the bulkheads. The inside water is crystal clear, but silt is stirred up as you pass through the 10-foot-high area.

Fifty feet north of the barge are the remains of the 91-foot wooden sailing ship, *Lady Free*.

NO.28 *ULTRA FREEZE*   14211.1   62150.0

Location: In the northern section of the R.J. Ventures Reef Site.

A 195-foot steel freighter sits at a depth of 120 feet. Her wheelhouse begins at 70 feet. The wreck has been down since 1984 and has attracted a fantastic fish population.

NO.29 *TARPOON* / KEVORKIAN MEMORIAL REEF   14210.2   62151.4
Location: In the R.J. Ventures Site, 500 yards inshore from the Ultra Freeze.

The 155-foot Haitian grain freighter, *Medor Herode*, was renamed the *Tarpoon* and placed in 71 feet of water as a tribute to Mike Kevorkian, owner of the Tarpoon Skin Diving Center in Hialeah since 1956. The riveted steel-hulled vessel, built in 1908, was declared unseaworthy by the Coast Guard. It is a beautiful vessel that sits upright, intact, with prop and rudder. The 30-foot profile makes an easy wreck dive for intermediate divers.

NO.30 HOPPER BARGE   14210.3   62149.9
Location: About 5 miles southwest of Key Biscayne in the center of the R.J. Diving Ventures Artificial Reef Site.

The 150-foot New York garbage barge sits on her side. She has a 40-foot-plus relief off the 163-foot bottom. This is a popular fishing site because of the concentration of large fish that always seem to be about. A huge school of jacks is never far away.

NO.31 *STEANE D'AURAY*   14209.6   62151.8
Location: Toward the southern end of the R.J. Ventures Artificial Reef Site.

A beautiful 110-foot North Atlantic trawler was sunk March 28, 1986 in 68 feet of water. She has a 28-foot profile. Locally she is called the St. Anne. This is an open wreck that is exciting to explore. It is a great swim-through ship with holes cut in the bulkheads. You can go down the hallway, past the galley and dining room, and on to the large engine room where you swim up to the pilothouse. The main deck hatch leads you to a large refrigeration compartment; then you're through the bulkhead to the forward berth, and exit a hole in the bow. What a trip!

NO.32 FOWEY ROCKS
Location: 11-1/2 miles south of Government Cut, marked by a 110-foot lighthouse.

This is a large area, as shallow as 10 feet and dropping to depths in excess of 100 feet. There is an abundance of angelfish, soft corals and some elkhorn coral. The entire area is enjoyable. It is a good spot for a repetitive dive after exploring the deeper nearby wrecks. Fowey Rocks is the start of a broken, shallow reef line that runs south into John Pennekamp Coral Reef State Park. Most of the site shares the same topography and marine life. The charm of Fowey Rocks lies in the abundance of fish life, living coral gardens, clear water, their sheer size, and the fact that they are less often dived than the Keys' reefs just to the south. Much virgin bottom still remains to be discovered. This and all the following reefs are located in the protected boundaries of the Biscayne National Park. The eastern boundary of the park extends to 60-foot depths. Its southern parameters join those of Pennekamp Park. Fishing, lobstering, conching and spearfishing are allowed. However, collection of tropical fish and relics is not permitted. Stated simply, take nothing that can't be eaten.

NO. 33 SONAR STRUCTURE

Location: 500 yards due east of the Fowey Light.

This large dish sits on a natural ledge in 80 feet of water. The entire area is brimming with sea life.

NO.34 BREWSTER REEF

Location: 2 miles south of the 110-foot steel freighter on Fowey Rocks. The reef is less than one mile west of the Blue Fire wreck site.

This is another large reef area with depths from 25 to 75 feet. Scattered coral heads, soft corals and sponges dot the sand floor. Visibility is usually good, but is best at high tide. Some scattered wreckage can be found along the blue outside edge of the reef in 70 feet of water. This is a good location for the experienced as well as the novice diver.

NO.35 *BLUE FIRE*   14204.6   62155.5

Location: West of Brewster Reef, in the North Dade Sportfishing Council Site.

The *Blue Fire* is a 175-foot freighter seized by the Coast Guard during the Mariel boat lift. She was sunk in 110 feet of water in January 1983. Her relief is over 30 feet off a beautiful white sand bottom covered with soft corals and large sponges. Intact and bearing full superstructure, except mast, this is indeed a wreck diver's wreck. It is easy and safe to penetrate her inner structure. The site has become home to a large number of midnight parrotfish, jacks and cobia. Don't miss this one!

NO.36 STAR REEF

Location: 4-1/2 miles south of the 110-foot steel lighthouse on Fowey Rocks and slightly east of the general reef line.

Star is considered to be a favorite reef by many Miami divers. Its rugged bottom is always teeming with marine life. It seems that something unusual can always be found among the coral heads. Depths can be as shallow as 12 feet.

NO.37 MYSTERY REEF

Location: A small reef area situated halfway between Star and Triumph Reef, 2 miles west of the Biscayne Monument's North Marker.

This is a beautiful patch reef area with magnificent stands of elkhorn coral. Schooling fish are dominant. There are also many large puffers in the area.

NO.38 TRIUMPH REEF

Location: 8 miles south of the 110-foot steel lighthouse on Fowey Rocks, on the same reef as Brewster, Star and Mystery Reefs. There is a red marker near the outside edge of the reef.

This is another patch reef of scattered coral heads. Depths on top of the reef are only ten feet. The best diving is in the 40- to 60-foot range, but good diving can be found as deep as 120 feet. Because of exceptionally clear water and rugged underwater terrain, this has long been a popular site.

NO.39 *ALMIRANTE*   14187.5   62173.8

Location: Found in the South Dade Sportfishing Council Site, halfway between Long and Ajax Reefs.

This is a 210-foot island freighter that sat for years in the Miami River. She was sunk in April of 1974 in 122 to 132 feet of water. She sits upright with her bow facing southeast. It is a 110-foot dive to her main deck. Because of her long stay underwater, her structure has attracted a wide variety of marine life. Her sides are covered with brilliant red soft coral and graced by schooling fish, creating a beautiful seascape for underwater photography.

NO.40 LONG REEF

Location: 10 miles south of the 110-foot steel lighthouse on Fowey Rocks, less than a mile west of the Almirante wreck.

This is a large reef area, running south for over two miles. There is a lot of fish activity here. This site has long been used for checkout dives because of its gently sloping bottom, making it ideal for multi-level dives.

NO.41 AJAX REEF

Location: Ajax Reef and her sister reef, Pacific, are the southernmost reefs in the Miami area. Ajax is actually a continuation of Long Reef. It is about 13 miles south of the Fowey Rocks lighthouse.

These are the last reefs in the Biscayne National Park area. Both have vivid coral structures and offer diving ranging from snorkeling depths to over 100 feet. The continuation of the reef line takes you into Pennekamp Park. The boundary line between the Biscayne National Park and Pennekamp is prominently marked by three pilings displaying warning signs. Of course, there is no spearfishing allowed within the Pennekamp boundaries.

NO.42 DOC DEMILLY MEMORIAL REEF   14181.7   62180.4

Location: In the Fish and Game Unlimited Artificial Reef Site, the southernmost of Miami's artificial reefs, 1/4 mile east of the Pacific Reef Lighthouse.

This is a large ,287-foot steel freighter sitting in 140 feet of water. It was sunk on March 6, 1986 as a memorial to a Homestead veterinarian. She sits upright with a 70-foot profile.

Inshore are two F-4 Phantom jet fuselages in 80 feet of water. They are 50 feet apart **14181.4 62181.7.**

27
KEY LARGO
NATIONAL MARINE SANCTUARY
JOHN PENNEKAMP
CORAL REEF STATE PARK

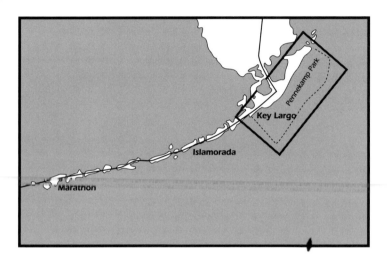

Offshore Diving Locations

1. Carysfort Reef
2. Carysfort South
3. The Elbow
4. North Dry Rocks
   Minnow Cave
5. Christ of the Deep—
   Key Largo Dry Rocks
6. Grecian Rocks

7. French Reef
8. Molasses Reef
9. *Thiorva*
10. HMS *Winchester*
11. *Towanda*
12. *Benwood*
13. Windlass Wreck
14. *Duane* and *Bibb*

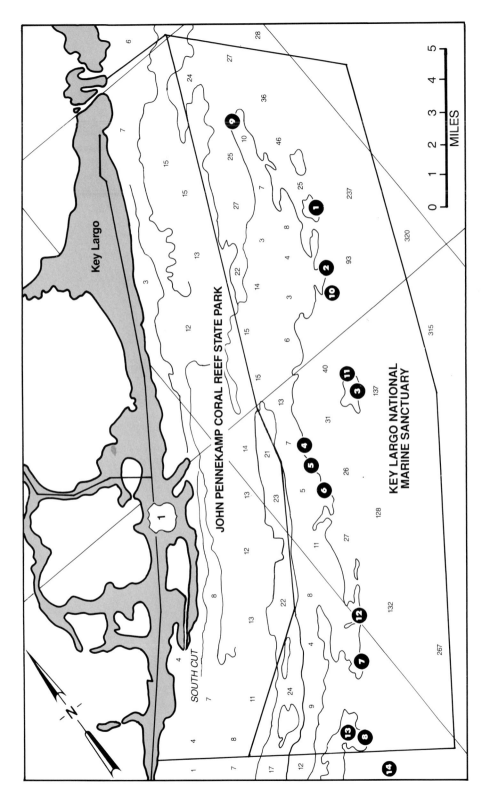

Area Information

In the Atlantic waters southwest of Key Largo are the most extensive coral reef systems in the United States. The massive coral banks are on the northernmost fringe where coral reef development is possible. To help conserve this vulnerable ecosystem, two government agencies have established marine preserves off Key Largo.

The John Pennekamp Coral Reef State Park was founded on December 10, 1960 after a strong lobbying effort by Florida conservationists led by Miami newspaperman, John Pennekamp. The State Park's boundaries encompass 75 square miles of ocean, extending from Key Largo to the end of the state's jurisdiction, three miles offshore. This area is maintained by the Florida State Park Service and protected by both the Department of Natural Resources and Marine Patrol. The park's Visitor Center is located at mile marker 102.5 in Key Largo and is open seven days a week.

On December 18, 1975, the Key Largo National Marine Sanctuary was designated by the Secretary of Commerce. For the first ten years, however, the state supervised the sanctuary for the federal government. In 1984, the National Oceanic and Atmospheric Administration established offices in Key Largo and began an active management role. The sanctuary shares Pennekamp Park's western boundary, three miles from shore, and extends seaward to the 300-foot isobath on the continental shelf. Their jurisdiction encloses a 104 square mile area, three to six miles wide, and 20 miles long. Its northern perimeter is the Biscayne National Park and extends south to Molasses Reef. The 14 miles of coral reef banks all lie within the sanctuary's boundaries, so the vast majority of sport diving activities take place in their jurisdiction.

Most of the sanctuary's bank reefs rise 10 to 15 feet, while some loom over 25 feet off the ocean floor. Each has a shape and personality of its own and stands cloaked in a living sea life tapestry. These world-famous reef communities are formed by over 52 species of West Indies corals and are home to 500 fish species. Every crevice, cave, overhang and outcropping swirls with life. Sea fans, rods and whips sway in gentle life-sustaining currents, while curtains of schooling grunts, tangs and yellowtails sweep by on every side. A day swimming over these reefs is a day never forgotten.

Unfortunately, human impact on the fragile coral communities has both short and long term negative effects. To help lessen stress on the reefs, the sanctuary staff has established a comprehensive management plan that includes education, regulations, research opportunities, and an extensive mooring buoy system.

EDUCATION

The State Park and Marine Sanctuary both encourage sport diving activities, but strive to minimize damage from the thousands of visitors. Their primary goal is to educate the diving public about the reef's fragile nature. Few divers realize that coral is a living animal and even minor contact with the coral, whether by scraping, rubbing, or standing on the structure, can cause severe damage to the tiny coral polyps. Each injured area can set the stage for a fatal infection from blue-green algae.

The role education can play in the reef's survival was well stated by former Sanctuary Manager Bill Harrigan.

"Everything in the coral reef environment is important to the health of the reef. The hard and soft corals, algae, coral rubble, sand, invertebrates, fish and sea grasses play vital parts. We, the visiting divers, are the only element not vital to the life of the reefs. If we expect to enjoy these beautiful reefs this year and the years to follow, then knowledge of the coral and respect for the reefs must be taught to each diver at the beginning of his training."

RESEARCH

Marine research, especially that related to the coral reef ecosystem and the impact of area visitors, is of primary importance. Scientific studies, including biological inventories, reef health assessments, deep water surveys, water quality assessments, and anchor damage and coral disease studies are ongoing projects within the sanctuary.

MOORING BUOY SYSTEM

A serious aftermath from so many visitors to the reefs is the damage caused by boat anchors. To help reduce this problem, mooring buoys were installed at frequently dived reef sites. The program evolved from the positive results attained after six experimental buoys were installed in 1981. An extensive mooring buoy program got underway in 1986. The buoys are available to everyone, on a first-come basis. Buoys also define reef areas, helping to prevent boat groundings. Soon, a complete system of over 200 should be in use. Each floating ball has one pickup line and is marked by a letter designator and numeral. The following is a list of buoys placed in 1988.

Location	Letter Designator	# of Buoys
Three Sisters	T	2
Molasses Reef	M	28
Sand Island	S	3
White Bank Dry Rocks	W	5
French Reef	F	17
Benwood Wreck	B	4
Grecian Rocks	G	12
Key Largo Dry Rocks	D	16
North Dry Rocks	N	2
North North Dry Rocks	NN	3
The Elbow	E	9
Carysfort South	CS	6
Carysfort Reef	C	11
Turtle Rocks	T	3
Turtle Shoals	T	7
Northeast Patch	NE	1
		Total 129

REGULATIONS

The quantity of sea life that prospers on Key Largo's reef systems is due, in great part, to the protective laws that have been enforced since 1960. The regulations are intended to protect the local ecosystem while providing maximum use by the public.

Regulations forbid:

- handling or standing on coral formations
- use of spearguns or wire fish traps
- removal or destruction of natural features, marine life (except the taking of spiny lobster and stone crab within season) and archeological and historical resources
- tampering with markers, mooring buoys and scientific equipment
- damage caused by boat anchors. No rope, wire anchor, or other shall be attached to any coral, rock or other formations.

Fishing, both sport and commercial, with hook and line is permitted.

These regulations are enforced by the Coast Guard and sanctuary Officers. Violators are subject to civil penalties of up to $50,000 under Public Law 92-532. A minimum fine of $150 has been established for minor anchor damage. Divers will only begin to appreciate the implementation of these rules after observing, firsthand, the abundance of protected marine life thriving within the boundaries.

NO.1 CARYSFORT REEF   14160.4   62211.9

Location: About 12 miles northeast of the South Sound Creek on the outside edge of the sanctuary. The reef is plainly marked by a 100-foot steel light tower.
Mooring Buoy System: C1-6 are on the seaward side of the elkhorn garden. C7 to C11 are on the shore side of the light, behind a shallow reef.

Carysfort is one of the outstanding dives in the marine sanctuary. The main reef starts with an elkhorn reef crest often exposed during low tide. Sloping seaward is a soft coral community that joins a staghorn coral forest in 42 feet of water.

Located east-northeast of the tower is "Carysfort Wall," a beautiful cascading wall of staghorn coral dotted by heads of brain and sheet corals. The wall drops 65 feet to a sandy bottom, and is a good dive for those who like a little more depth.

Four hundred yards north-northwest of the tower, in 25 to 40 feet of water, are two large anchors, possibly from an old sailing frigate of the 1800s. This extensive reef area is thought to be the final resting place of the 1755 Spanish Plata Flotilla. A hurricane drove the fleet onto the reef, but no trace has yet been found of the hapless vessels.

Carysfort Reef Lighthouse was the first lighthouse built on exposed reef in the Keys. Construction began in 1848. Its light was ignited in 1852. NOAA is planning to renovate the structure so that it can be used as a base of operations for visiting scientists.

NO.2 CARYSFORT SOUTH   14158.3   62214.8

Location: The southern reef extends two miles south from the Carysfort tower.
Mooring Buoy System: CS1-4 are on the reef's seaward side, CS5-CS6 are on the shore side.

This long, narrow reef is one of the most striking elkhorn gardens in the sanctuary. It runs for several hundred yards. Depths vary from two to 20 feet. Several canyons wind through the reef. Large schools of blue tangs and grunts are common and can be easily approached by the diver. The deep water reef is completely isolated from the shallow reef. It is one of the only deep areas where staghorn coral is present.

Dive Pennekamp
WITH
DIVERS DEN

HOTEL DIVE PACKAGES

Wreck, Reef, Night, Photo
"Fish Bowl of the Caribbean"

DIVE TRIPS

•Scubapro • U.S. Divers • Dacor • Tabata
"All Major Brands"

SALES & SERVICE

PADI 5 ★ Training Facility
"Open Water Diver through Assistant Instructor"

INSTRUCTION

• SPECIALTY COURSES AVAILABLE
• RESORT COURSE AVAILABLE

RENTALS & AIR

• SIX PACK CHARTERS AVAILABLE

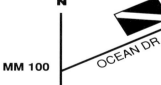

N

MM 100

OCEAN DR.

KEY LARGO

SHERATON
KEY LARGO RESORT

MM 97

S

DIVERS DEN	**DIVERS DEN**	**DIVERS DEN**
12614 N. Kendall Dr.	110 Ocean Drive	2333 S. Univ. Dr.
Miami, FL 33186	Key Largo, FL 33037	Ft. Laud., FL 33324
(305) 595 - DIVE	**(305) 451- DIVE**	**(305) 473 - 9455**

NO.3 THE ELBOW   14148.6   62229.3

Location: About 8 miles east-northeast of the South Cut, or 5-1/2 miles southeast of the North Cut off Key Largo. The area is marked by a 36-foot light tower with #6 on it.
Mooring Buoy System: Marked by nine buoys, E1-9.

The Elbow, or the "wreck reef" as it is commonly called, is littered with the bones of various ships. One hundred fifty yards east-northeast of the tower, in 20 feet of water, lies the scattered wreckage of the *City of Washington*, a large steel freighter that went aground in 1791. This is a beautiful snorkeling spot. Twenty yards east-southeast of the tower,lie the broken remains of a steel barge called Mike's Wreck. She is in 18 to 25 feet of water (between E2 and E3).

About 80 yards west of the steel ship lies the "Civil War Wreck" consisting of large piles of timbers with bronze fastening pins still visible. Beneath E4 among the coral ridges is a length of old chain, possibly off the *Towanda* (see No.11). A six-foot canyon from the 17th or 18th century lies encrusted with coral approximately 100 yards west of this wreck.

Fifty yards east-southeast of the tower are beautiful finger reefs of elkhorn coral, ranging in depths from six to 20 feet. Here, hundreds of damselfish and angelfish make their home.

NO.4 NORTH DRY ROCKS MINNOW CAVE   14147.0   62240.0

Location: 1/2 mile northeast of the Key Largo Dry Rocks.
Mooring Buoy System: The reef stretches between N1 and N2.

This is a nice reef area with depths to 15 feet. In the central portion of the small reef is the Minnow Cave. A swirl of glass minnows constantly adorn the opening to the narrow coral cave. This is a good, out-of-the-way snorkeling spot.

NO.5 CHRIST OF THE DEEP (KEY LARGO DRY ROCKS)   14143.0   62241.4

Location: About 6 miles east-northeast of the South Cut on Key Largo, and marked by an orange and white surface buoy.
Mooring Buoy System: Sixteen buoys, D1-16, outline the reef area.

Christ of the Deep, also referred to as Christ of the Abyss, is a replica of "Il Cristo Degli Abssi," located in the Mediterranean Sea near Genoa. The 4,000-pound, nine-foot statue of Christ with his arms raised upward was created by Italian sculptor Guido Galletti. In 1961, another statue was cast from the same mold for Edidi Cressi, a European industrialist who presented it as a gift to the Underwater Society of America. After much deliberation, the Society decided to place the statue in the clear water of the Florida Keys. Today, the statue graces the Key Largo National Marine Sanctuary, where thousands of divers visit it each year. It sits on a concrete base in 25 feet of water.

About 300 yards around the edge of the reef, to the northwest (inshore side), the water is 20 to 25 feet deep and calmer in rough weather. The area is covered with a large variety of tropicals. Skates and leopard rays are commonly sighted in the white sand around the coral.

The statue is surrounded by coral cliffs and several huge heads of brain coral. This is a dramatic site for the underwater photographer. Snorkeling is good behind the statue, with an average depth of five to 10 feet.

Christ of the Deep.   Richard Collins

NO.6 GRECIAN ROCKS   14141.2   62243.6

Location: About 3/4 of a mile south-southwest of Christ of the Deep. The area is marked by G1- G12.

Mooring Buoy System: G1-12 outline the reef area. During low tide, part of the area is awash.

Grecian Rocks has a well-defined reef crest of densely packed elkhorn coral. This is one of the most popular snorkeling areas in the sanctuary. The reef is half a mile long (running north to south) by 150 yards wide. A fore reef slopes to a white sand plateau. Tropical fish are abundant here. Many are accustomed to being hand-fed. A Spanish cannon is located 75 feet south. A cluster of star coral is located on the shoreward side of the reef. Depths vary from four to 25 feet.

NO.7 FRENCH REEF   14129.6   62260.2

Location: About 4-1/2 miles south-southeast of South Cut off Key Largo, or about 1 mile northeast of the steel tower on Molasses Reef. The reef area is marked by a black piling.

Mooring Buoy System: Buoys F1-17 mark specific dive sites.

This is one of the most dramatic reef areas in the Florida Keys. Limestone cliffs are mantled with hard and soft corals and honeycombed with dozens of tunnels, caves and crevices. Many are large enough for divers to swim through easily. Exploring with an underwater light is recommended. The beam will help you discover a multitude of sea life hidden in the dark recesses. Schooling fish, including yellowtails, grunts and porkfish, sweep endlessly through the crevices while the caves teem with copper sweepers. Of course, morays love all the groovy hiding places.

Just northeast of F1 is a 100-foot undercut that runs toward the sea. There are two swim-throughs in the ledge. The Hourglass Cave is 50 feet inshore, across a sand bed.

The Christmas Tree Cave is 50 feet inshore from F3. Spiraling Christmas Tree worms, commonly found on the corals, give the 20-foot tunnel its name. Grouper and morays love this spot.

One hundred feet west from F5 is a round sand bank bordered by a coral ridge. On the southeast side is a low shelf-cave with a white sand floor that can be traversed seaward.

Another swim-through is found on the seaward end of a coral ridge 25 feet inshore from F6. Just inshore from F7 is a coral-covered ledge that runs northeast and southwest. Follow the ridge 100 feet northwest and you will find an old anchor entombed in the coral.

NO.8 MOLASSES REEF   14124.9   62268.5

Location: 5 miles south of the South Cut off Key Largo at the southern end of the sanctuary. It is marked by a 45-foot steel light tower. The best diving is on the southeast side of the tower marked by buoys M1-M8.

Mooring Buoy System: M1-25 mark sites suitable for diving or snorkeling. M21-23 are deeper diving sites.

Molasses is the most popular diving area in the sanctuary. On any given weekend the site is carpeted with private boats and charter craft. What attracts so many is the large barrier coral community that swarms with sea life. Depths to the white

sand between the high-profile reef sections average 15 to 40 feet. Fish watchers will have a field day trying to count the many species that inhabit the reef. Commonly sighted on every dive are barracuda, yellowtail, sergeant majors, porkfish, grouper, snapper, chub, blue tang and angelfish galore.

Because of the reef's close proximity to the Gulf Stream, visibility is often better than other sites. The reef's northern end is shallow and best for snorkeling, while deeper dives to 80 feet can be made in the southern section. The short boat trip from Key Largo and the ease of finding the reef make it a natural for night diving.

Adjacent to M3 is an eight-foot Spanish anchor. "Hole in the Wall" is a swim-through found just southwest of M8. The deeper reef section, to 55 feet, is marked by M21-23.

In 1984, the ocean-going freighter *Wellwood* ran aground on the reef and remained fast for several days. Damage was considerable. The hull plowed through the coral and deep into the substrata. Buoys M11-12 are next to the damaged site. Please stay clear of the rubble area so that a new coral community will have a chance to develop.

NO.9 *THIORVA*   14166.4   62206.5

Location: North end of Turtle Reef.
Mooring Buoy System: None
The wreckage lies scattered among the coral. Nothing of value has been discovered in the area, but wreckage remains encrusted in the coral and buried beneath the sand.

NO.10 HMS *WINCHESTER*   14156.8   62216.5

Location: British Ship of the Line
Date: 1695
Location: 1.5 miles southwest of Carysfort Light, on a direct line between Carysfort and the Elbow, in 28 feet of water.
Mooring Buoy System: None.

Much of the *Winchester* is encrusted with coral, making it hard to locate. Broken pieces of the wreckage can be found over a large area. The 140-foot British frigate was lost in a violent storm in 1695. Most of her sixty guns were taken for scrap iron during WWII. Two of her cannons, recovered in 1940, are on display at the Coral Reef State Park headquarters.

NO.11 *TOWANDA*

Class: Steamer
Date: 1866
Location: Just north from the Elbow Reef tower.
Mooring Buoy System: E4 is above a length of old chain. E5 marks the wreck site.

Half buried remains of the steel-hulled vessel lie among the coral heads.

NO.12 *BENWOOD*   14132.9   62254.3

Class: Freighter
Date: 1942
Location: From the pile on French Reef, travel northwest toward a Red Nun buoy (difficult to see on the horizon) for approximately 1-1/2 miles. The ship lies almost halfway between and on a direct line with a pile and buoy.
Mooring Buoy System: B1-4.

During World War II, the freighter M.V. *Benwood* was torpedoed off the Florida Keys by a German submarine. As she limped toward shallow water to go aground for easy salvage, she was rammed by a friendly ship. Later, five shells exploded amidship, finishing her off and sending her to the bottom. Her hull was used for bombing practice, and the bow was dynamited because it was a hazard to navigation.

The *Benwood* is 285-feet long and lies in a northeast line. Her bow is in 25 feet of water, and the depth increases to 55 feet at the stern. An eerie underwater setting is created by the huge aft section that looms sharply upward off the ocean floor. The maze of steel wreckage provides a haven for large numbers of fish, from small tropicals to large grouper and snook.

NO.13 WINDLASS WRECK

Class: Schooner
Date: Unknown
Location: 30 yards due east of Molasses tower in 25 feet of water.
Mooring Buoy System: M8.

Broken remains, including a large anchor windlass, lie buried in the sand between coral heads.

NO.14 *DUANE* AND BIBB   14122.2   62270.9 &   14122.9   62270.3

Class: Coast Guard cutter
Date: November 27, 1987 (sunk to become artificial reefs)
Location: The Duane lies approximately one mile south of the Molasses light tower. The Bibb is 1/2 mile north of the Duane.

An added attraction to the superb reef diving off Key Largo are the two beautiful 327-foot Coast Guard cutters that rest on a white sand bottom just outside the sanctuary's southwest boundary. The cutters were intentionally scuttled on November 27, 1987 to become artificial reef sites.

Two safety factors must be taken into consideration when planning a dive on these vessels. The first is depth. To keep their huge hulls from becoming hazards to navigation, both vessels were placed in relatively deep water. The *Duane* rests upright with a slight starboard list. She faces southwest. Her deck is 90 feet deep with a crow's nest extending to within 50 feet of the surface. The *Bibb* is in 130 feet of water, lying on her starboard side with her bow facing north. It requires a 95-foot dive to reach her port gunnel railing.

Currents sometimes create problems. Because the ships are further to sea than the reefs, the Gulf Stream has a greater influence here. Generally, the current will be of little concern, but on certain days the water can rip with such force that safe diving is impossible. Always descend on the anchor or mooring line and explore against the flow.

The intact vessels were carefully prepared to accommodate divers. Superstructures were left open for exploration. An underwater light is

Paul Humann

recommended when touring the interior. The hull has been closed off for safety. Visibility can exceed 100 feet, offering a spectacular view of the steel monoliths. This is a must for advanced divers!

28

UPPER KEYS

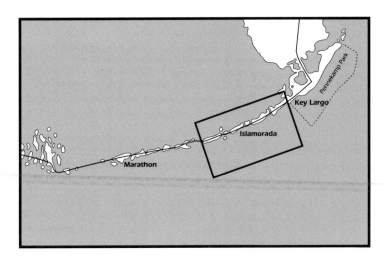

Offshore Diving Locations

1. Conch Reef
2. Little Conch Reef
3. Davis Reef
4. Crocker Valley
5. Crocker Reef
6. Crocker Wall
7. Hens & Chickens Reef
8. Rocks
9. Cheeca Rocks
10. Mini No Name Reef
11. Hammerhead Reef
12. Canyon
13. Alligator Reef
14. Alligator Gulley
15. Matecumbe Drop-off
16. Caloosa Rocks
17. *Eagle*

18. *Alexander* Barge and Artificial Reef Site
19. *D & B* Wreck
20. Cannabis Cruiser
21. Spiny Oyster Barge
22. Iron Masted Schooner
23. USS *Alligator*
24. Brick Barge
25. *Infante*
26. *San Jose*
27. *El Capitan*
28. *Chaves*
29. *Herrera*
30. *Tres Puentes*
31. *San Pedro*
32. *Lerri*

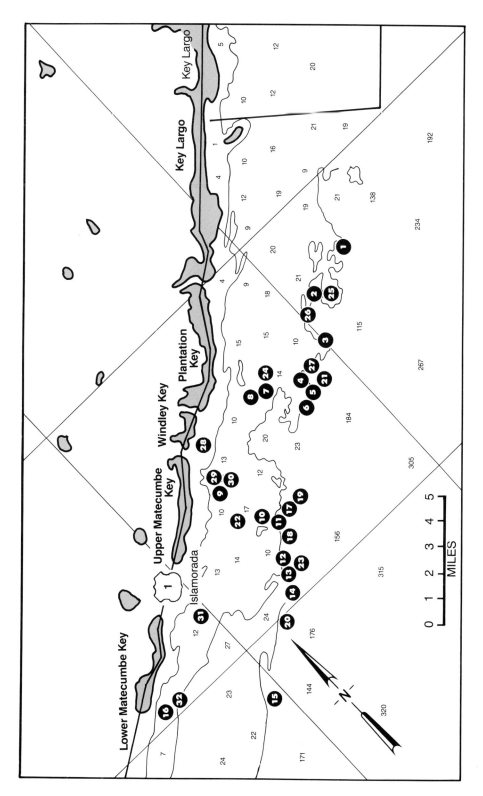

Area Information

Diving experiences in the Upper Keys range from snorkeling over exquisite coral gardens to plunging down the sides of underwater drop-offs. The sunken remains of some of the richest treasure galleons ever discovered rest in these waters. In past years, the huge piles of egg-shaped ballast stones have given up thousands of silver coins and rare artifacts to hardworking treasure hunters. Much treasure still lies hidden under the sand and in acres of grass beds. Adding to the romance of these old galleons is a collection of modern vessels that now grace the sea floor just off Islamorada. Several have been down for years attracting a multitude of sea creatures. Two large ships have recently been sunk to create exciting artificial reef sites.

Diving guide services to the Upper Keys reefs and wrecks are found in Tavernier and Islamorada. There are many resorts and motels that cater to divers.

As these waters are outside the Pennekamp Park boundaries, spearfishing is permitted here. Several charter boats allow spear guns on board. If you plan to spear, please stay away from the sport diving sites mentioned in this chapter. The best locations for underwater hunting are those seldom frequented by divers.

NO.1 CONCH REEF
Location: 8 miles east of Windley Key, marked by two piles.

A rugged, sloping terrain with depths ranging from 15 to 20 feet, this area is full of interesting coral formations and sea fans. Conch Reef is known for its outstanding visibility year round, and its great mounds and valleys. This reef abounds with both large and small fish, and is excellent for the collector. The area is bordered by thousands of conch shells, from which the name of the reef is taken.

NO.2 LITTLE CONCH REEF
Location: 6 miles east of Windley Key.

A large patch reef area spreads for 1/2 mile across the white, flat sand bed just inside the inner reef. Depths average 30 feet. Sea fans and tube sponge colonies are common. There are a few stands of elkhorn coral scattered about. This is a great place to bag lobster during the season. The Little Conch Ledges are located 100 yards west. The Infante lies just 75 yards southwest.

NO.3 DAVIS REEF
Location: 3 miles east-southeast of Hens and Chickens Tower.

Davis is an outer reef known for its giant schools of fish. Literally, "schools of thousands" of tiny fish, all moving in unison to the rhythm of the sea, accent this beautiful reef. Striking stands of coral are covered with a multitude of sponges and sea fans. Depth averages 35 feet.

NO.4 CROCKER VALLEY
Location: Just outside Crocker Reef.

A river of white sand 60 feet wide spills down the outer reef's slope. It starts at the reef's crest in 45 feet and runs down the incline to 110 feet. Vertical coral walls 20 feet high border both sides of the sand valley. The sea creatures that roam here are a bit larger and wilder than those commonly found on the inner reefs. Nurse sharks, sea turtles and amberjacks are nearly always sighted during a swim down the slope.

NO.5 CROCKER REEF
Location: 5-1/2 miles southeast of Windley Key.

One of the outer reefs, Crocker ranges from 35 to 100 feet deep and is graced with beautiful formations of staghorn coral and swaying gorgonians. This is a paradise for the tropical fish collector, with an array of multicolored sponges and varying sea life.

NO.6 CROCKER WALL
Location: On the south side of Crocker Reef.

This is another sea slope dive. An array of waving sea fans and gorgonians cover the coral-encrusted incline. Large barrel sponges are common. Rays, turtles and grouper inhabit all levels of the reef.

NO.7 HENS & CHICKENS REEF
Location: 3 miles east-southeast of Windley Key. Marked by a 35-foot tower.

Hens and Chickens Reef abounds with massive boulders of brain coral which reach up to 20 feet in height. Here, the underwater photographer can find settings of iridescent tropical fish swimming and multicolored gorgonians.

At the north end of the reef lies the wreck of the "Brick Barge," an old steel barge that went down after being torpedoed during World War II. Depths range from 20 to 28 feet.

NO.8 ROCKS
Location: 2-1/2 miles due east of Windley Key, about 3/4 of a mile off Plantation Point.

A beautiful underwater coral garden that provides a wonderful sight for snorkelers in water just 8 to 12 feet deep. The bottom is comprised of soft coral and sea fans. This is a good area for shelling, tropical fish collecting, lobstering or just plain exploring.

NO.9 CHEECA ROCKS
Location: 3 miles south-southwest of Windley Key, or 1 mile directly off Upper Matecumbe.

A picturesque setting for the photographer, these coral gardens contain large mounds of brain coral and a wide variety of soft corals. An excellent reef for snorkeling, and a good place for beginners, this reef averages 12 feet in depth.

NO.10 MINI NO NAME REEF
Location: 4 miles south of Windley Key.

This is a beautiful reef section on the outer fringes of No Name Reef. Depths vary from 35 to 50 feet. Narrow sand trenches wind through stands of sheet and flower coral. Mangrove snapper and hog fish are abundant.

NO.11 HAMMERHEAD REEF
Location: Just south of Mini No Name Reef.

A section of the outer reef that gradually slopes from 50 to 65 feet before plunging to 95 feet.

NO.12 CANYON
Location: One mile north of Alligator Reef.

This is a dramatic section of outer reef that, because of the depths, is recommended for advanced divers. The reef starts at 65 feet and drops to 115 feet.

NO.13 ALLIGATOR REEF
Location: 6 miles south-southwest of Windley Key. Marked by a 136-foot tower.

Named after the USS *Alligator*, a U.S. warship sunk in 1825, this reef is one of the largest in the Upper Keys. The deep coral crevices and ravines have made this a popular area for underwater exploration. Here, the diver will find all types of coral, both hard and soft, as well as numerous tropical fish and shells. The broken remains of the warship can be found a few hundred feet on the ocean side of the light. Depths range from 8 to 40 feet.

NO.14 ALLIGATOR GULLY
Location: 300 yards west-southwest of Alligator Reef.

If you like to dive where there is an abundance of marine life, the Gully is your spot. Coral-studded ledges 8 to 28 feet high rise to near the surface. Sweeping through the coral valleys are legions of schooling fish. Chubs, tangs, porkfish and mutton snapper are here by the hundreds. Tarpon, snook, grouper and barracuda are also commonly sighted.

NO.15 MATECUMBE DROP-OFF
Location: Located between, and on the line with, the 136-foot tower on Alligator Reef and the 49-foot tower on Tennessee Reef.

The good areas on this drop-off are hard to find without a guide or depth sounder. This area provides a lot of excitement for the more experienced diver. The drop-offs vary in depths, but usually run from 35 to 135 feet. The greater depths offer a variety of marine life unlike that found in the shallows.

NO.16 CALOOSA ROCKS
Location: 1 mile directly offshore from Caloosa Cove Marina.

The Coral Gardens is an impressive dive with its many stands of brain, lettuce and star corals. Underwater photography is excellent and the reef fish are friendly. Depths vary from 5 to 18 feet, providing good snorkeling and shallow scuba. This is also an excellent area for night diving.

NO.17 *EAGLE*   14093.2   43292.9
Class: Freighter
Date: Modern
Location: 4 miles southeast of Islamorada.

On December 19, 1985, the 287-foot freighter *Eagle* was sunk in 110 feet of water

The Eagle going down.

to create an exciting wreck dive for the thousands who visit offshore Islamorada each year. The combined efforts of local diving businesses are responsible for the expensive, time-consuming project.

The *Eagle* settled on her starboard side with the bow pointing north. The entire superstructure with a crow's nest and cargo booms remained intact. Because her profile rises over 40 feet off the ocean bed, it takes only a 65-foot dive to reach the vessel. Within months after her sinking, a myriad of schooling fish took up residence in her protective confines.

Light streams inside the great structure through the two large cargo holes, portholes and the eight point where her hull was blown open during the sinking.

NO SPEARFISHING IS ALLOWED ON THE *EAGLE*.

NO.18 *ALEXANDER* BARGE AND ARTIFICIAL REEF   14093.0   43293.2

Class: Barge
Date: Modern
Location: Just west of the Eagle.

The 120-foot long, 40-foot wide barge Alexander was sunk in 1984 to initiate the development of an artificial reef site. The vessel lies intact in 105 feet of water. Rubble from the old Whale Harbor Bridge was dropped in the area the next fall. A great variety of marine life already have been attracted to the wreckage. Schools of snapper, grunts and jacks as well as porgies, hogfish and grouper abound. Infant soft and hard coral have begun to develop, providing sanctuary for numerous reef fish.

NO.19 *D & B* WRECK

Class: Barge
Date: Modern
Location: Less than 1 mile east of the Eagle.

This recently discovered liquid cargo barge appears to have been down for many years. Because of the time spent on the sea floor, the wreck is now a spectacular fish haven. A large green moray is a popular attraction. Grouper, mutton snapper and hogfish are always present.

NO.20 CANNABIS CRUISER   13927.1   43796.7
Class: Trawler
Date: Modern
Location: 3 miles west of Alligator Light.

The ill-fated efforts of drug pirates is now on display in 110 feet of water. Detection of their clandestine mission by the Coast Guard forced the hapless crew to scuttle their vessel with the cargo still aboard. She sits upright, intact, and is an adventure to explore.

NO.21 SPINY OYSTER BARGE
Class: Barge
Date: Modern
Location: West of Crocker Reef.

The barge lies in 105 feet on a flat seabed. Except for a large break in one side, the vessel is intact. Over the years the wreck has become home for a large colony of spiny oysters.

NO.22 IRON MASTED SCHOONER
Class: Schooner
Date: 1800s
Location: 2 miles south of Windley Key.

The remains of this 19th century sailing ship rest in 40 feet of water. Over the last century nature's forces have broken the ship apart. The large iron mast is now in three sections surrounded by mast hoops, steering machinery, a winch and anchor chain. Deadeyes, portholes, silverware and coins have been recovered by divers.

NO.23 USS *ALLIGATOR*
Class: Man of War
Date: 1822
Location: Between coral heads on the ocean side of Alligator Light.

In 1822, this ship was returning to Key West after a skirmish with pirates which left her commander, Lieutenant W.H. Allen, critically wounded. She now rests on the reef that bears her name.

NO.24 BRICK BARGE

Class: Barge
Date: Modern
Location: On the north end of Hens and Chickens Reef.
This is a very scenic steel wreck lying among picturesque coral heads.

Upper Keys Treasure Wrecks

On September 15, 1733, a strong hurricane engulfed the famous Spanish fleet as it skirted the Florida reefs riding the Gulf Stream toward Spain. Twenty-one galleons were driven over the dangerous reefs by the high winds and scattered from Key Largo down the chain of islands to Key Vaca. Today, large piles of ballast are all that remain to mark their final resting place.

The wrecks were first salvaged by the Spaniards. Centuries later, modern treasure hunters brought their airlifts and carefully combed the area of ballast for hidden silver and rare relics. However, much of the treasure is buried there today under the ever-shifting sands. The sport diver can still be rewarded by patiently fanning the sands around the wrecks. Small pieces of pewter and pottery are common, and an occasional piece of coral-encrusted silver is discovered.

NO.25 *INFANTE*   14109.0   43266.0

Class: Galleon
Date: September 15, 1733
Location: 75 yards southeast of Little Conch Reef in 25 feet of water.
A large pile of ballast stones mark this site. The wreck is known for unique silver coins with special edge markings called pillar dollars. Hundreds of them were recovered in the area. Supposedly, the Infante was carrying the first such coins minted in the New World.

NO.26 *SAN JOSE*   14108.5   43268.8

Class: Galleon
Date: September 15, 1733
Location: In 35 feet of water, about 1 mile east of Little Conch Reef.
The *San Jose* is still producing treasure. In March of 1973, a new section of the wreck, located not more than 50 yards from the main wreck, produced $30,000 in gold and silver on the first day of salvage. The ribs and keel are exposed, creating an exciting setting for underwater photographers.

NO.27 *EL CAPITAN*   14103.8   43276.5

Class: Galleon
Date: September 15, 1733

Location: In 20 feet of water, 6 miles east-southeast of Windley Key.

This wreck, marked by a huge ballast pile, has been the source of untold amounts of fascinating relics, as well as real treasure. Much remains!

NO.28 CHAVES   14098.5   43292.7
Class: Galleon
Date: September 15, 1733
Location: Just off Windley Key, south of the entrance to Snake Creek.

The Chaves rests in one of the many sand pockets in a grass bed in ten feet of water. This small ballast wreck still yields relics to patient divers.

NO.29 HERRERA
Class: Galleon
Date: September 15, 1733
Location: About 2-1/2 miles south of Snake Creek Bridge.

The ballast piles of Herrera lie in a grass bed in 18 feet of water. Many interesting relics have been recovered from this area, including small clay figurines of animals and fish.

NO.30 TRES PUENTES   14093.5   43296.5
Class: Galleon
Date: September 15, 1733
Location: In 20 feet of water, about 3-1/2 miles south-southeast of Snake Creek Bridge.

This area has produced quite a bit of silver. There is still some good relic hunting on the ocean side of the ballast pile.

NO.31 SAN PEDRO   14082.2   43320.8
Class: Galleon
Date: September 15, 1733
Location: This wreck lies on one of many sand pockets of a large grass bed about 1-1/4 miles due south of Indian Key.

Many coins have been recovered on the ocean side of this wreck.

NO.32 LERRI   14077.3   43330.6
Class: Galleon
Date: September 15, 1733
Location: 20 feet of water, 3/4 of a mile off Lower Matecumbe.

The huge 150-foot ballast pile hasn't produced much treasure, but many interesting relics have been recovered. This wreck provides a setting for some interesting photography.

29
MIDDLE KEYS
AND
LOOE KEY
National Marine Sanctuary

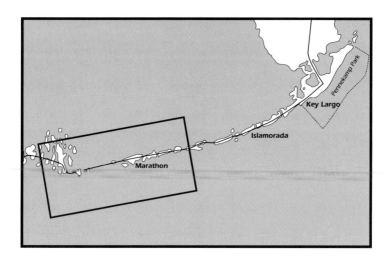

Offshore Diving Locations

1. Coffins Patch
2. East Washerwoman
3. Delta Shoals
4. Sombrero Reef
5. G Marker
6. *Ignacio*
7. Delta Shoal Wreck
8. Ivory Wreck

9. Fish Marker
10. Pillars
11. *Thunderbolt*
12. Yellow Rocks
13. Duck Key Wreck
14. HMS *Looe*
15. Looe Key
    National Marine
    Sanctuary

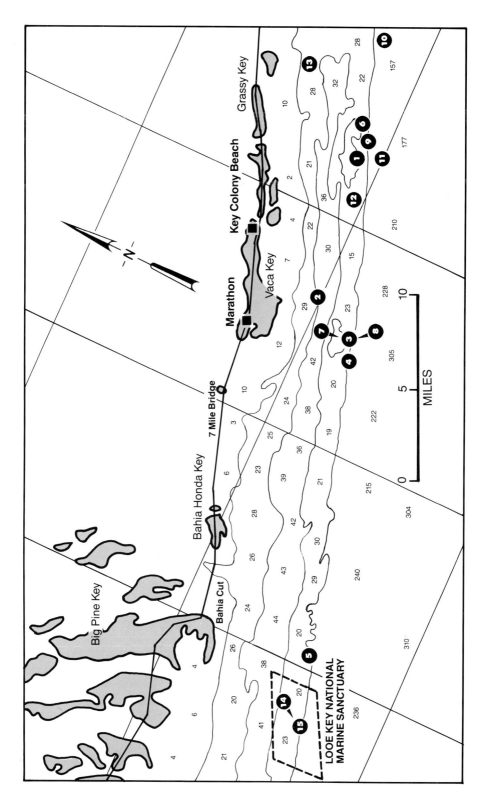

Area Information

The heart of the fabulous Florida Keys' diving lies in the Middle Keys. This is the location of Sombrero Reef and Looe Key, two of the most magnificent coral reefs in the Atlantic. Days could be spent exploring these reefs, but there is much more to do and see in the surrounding waters. Deep drop-offs bathed by the Gulf Stream challenge the experienced open-water divers, while acres of living reefs afford hours of shallow-water snorkeling enjoyment.

NO.1 COFFINS PATCH
Location: A large reef area running about 1-1/2 miles northeast and southwest, located 3-1/2 miles southeast of Key Colony Beach.

Coffins Patch offers several exciting scuba diving and snorkeling areas, with depths varying from 10 to 15 feet.

On the southwest end, a six-foot piling sticks out of the water. Anchor 50 yards east or west of the pile. This is a very colorful area that is a thrill to dive. Large 20-foot brain corals, fire coral and staghorn corals provide homes for an abundance of marine life. Spanish lobster and small spiny rock lobster can be found hiding under small ledges, and large schools of French grunts and mutton snapper are everywhere.

If you line up the pile with the Sombrero light and go 1 1/4 miles northeast, you will be over another beautiful reef with some of the largest heads of brain coral to be found in the Keys. This is a good spot for underwater photography, with many ledges alive with fish. Several large French angelfish will follow you around, waiting for a handout.

Continuing on the same line with the pile and light tower for another 200 yards, you will find another exceptional diving location covered with brain and staghorn coral. There is even more 300 yards to the south.

NO.2 EAST WASHERWOMAN
Location: About 6 miles southwest of Key Colony Beach.

Marked by a 36-foot tower, this is a good area for snorkeling, with depths from 10 to 18 feet. Many varieties of coral, including rose, midget brain, star and staghorn dot the bottom. Sea fans and sponges are common.

NO.3 DELTA SHOALS
Location: About 1-1/2 miles east-northeast of Sombrero light.

This area is easily sighted when approached. The shoal runs for half a mile, east and west, with depths from 10 to 20 feet. The best diving is on the scattered coral heads near the outside. The remains of several wrecks lie in the coral fingers on the ocean side of the shoal. The broken remains of an old slaver that went aground in the 1850s can be found buried in the sand between the coral fingers. She is called the "Ivory Wreck" because of the elephant tusks that have been recovered from her. No treasure was discovered on any of the wrecks, but many unique relics still remain hidden under the sand.

NO.4 SOMBRERO REEF

Location: About 8 miles southwest of Key Colony Beach. The reef is marked by a 142-foot light tower. The best diving is on the south side (ocean side) of the light. Anchor between the light and the coral heads.

The moment you leave your boat and start gliding down through the clear water surrounding the light, you'll realize the magnificence of Sombrero Reef. Several huge fingers of coral, covered with a stunning variety of gorgonians and brightly colored tropicals, run over the white sand bottom toward the ocean. A diver can easily use several tanks in the area and never fully explore all the ravines, ledges and archways formed by the large coral outcroppings. Cudas are large and plentiful and add to the bold setting created by the reef. Many varieties of coral are present, with large stands of brain coral and lettuce coral common.

The underwater photographer will need to load his camera with a 36-exposure roll instead of 20, because he will find many interesting settings bursting with nearly every color and shape a reef can display.

NO.5 G MARKER

Location: From the Bahia Honda Cut, go 5 miles south-southwest to a 36-foot tower.

This is a large, flat area with depths from 15 to 40 feet. Small scrub corals and sponge formations dot the white sand floor. Many large fish prowl the area, including plentiful schools of jacks. The area around the tower is a favorite haunt for large barracuda.

Beach camping in the Keys.   Ned DeLoach

NO.6 *IGNACIO*

Class: Galleon

Date: September 15, 1733

Location: 100 yards east-southeast of the old beacon on Coffins Patch Reef.

As the *Ignacio* was carried over the reef by high winds, her bottom broke open and her cargo, including much silver, was scattered for hundreds of yards over the flat grass bed.

NO.7 DELTA SHOAL WRECK

Class: Schooners and Galleons

Date: Unknown

Location: Between the coral fingers of Delta Shoals.

The broken wreckage of several ships lies hidden in the sand between the coral outcroppings. Remnants of both schooners and galleons have been recovered from this area.

NO.8 IVORY WRECK
Class: Schooner
Date: 1850s
Location: Between the coral fingers of Delta Shoals.

There isn't much to see of this slaver, which went aground on Delta Shoal in the 1850s, but many interesting relics have been recovered from the sands around her broken remains. Elephant tusks were found, helping to establish her as a slaver and giving the wreck the local name, "Ivory Wreck."

NO.9 FISH MARKET
Location: 4 miles southeast of Key Colony Beach.

This is a large, broken reef that runs for over a mile in 50 to 70 feet of water. The beautiful heads of living coral rise 15 to 20 feet from the white sand. Fish life, like the reef's name indicates, is the main attraction. Thousands of schooling fish roam through coral gardens, while larger grouper and hog snapper hide in the shallow undercuts. Lobster are common.

NO.10 PILLARS
Location: 8 miles east-southeast of Key Colony Beach.

Some of the Keys' tallest coral heads loom up 20 to 25 feet from the sea floor. Depths are from 60 to 90 feet. There are plenty of large fish around the beautiful coral spires.

NO.11 *THUNDERBOLT*   14034.0   43403.4
Class: Cable-layer
Date: Modern
Location: 1 mile south-southwest of Marker 20.

The clear waters off Marathon's outer reef have a new tradition to thrill divers. A 200-foot steel research vessel, the RV *Thunderbolt*, sits upright on the sand floor at a depth of 115 feet. She's a beauty. Her bridge comes within 75 feet of the surface. Doors and hatches were removed for the diver's safety, but nearly everything else remains intact, including two large bronze propellers.

The ship was built as a cable-layer during WWII, but was later converted into a research vessel for Florida Power & Light Co. She was sunk on March 3, 1986 to become the first of many artificial reef sites planned for the area.

NO.12 YELLOW ROCKS
Location: 3-1/2 miles south-southeast of Key Colony Beach.

This is a fine set of ledges that runs east and west for over 200 yards. They rise nearly 12 feet off the sea floor. Maximum depths are 22 feet. Undercuts and crevices conceal an interesting variety of marine life including nurse sharks and grouper.

NO.13 DUCK KEY WRECK
Class: Steamer
Date: 1800s
Location: 3-1/2 miles south of Duck Key.

The remains of a nameless steamer that went down around the turn of the century lie scattered in a nice patch reef area. Four large stacks, 50 feet long and 30 inches in diameter, are easily spotted. The maximum depth near the stacks is 25 feet. There is plenty of fish life around the large coral heads and broken wreckage.

NO.14 HMS *LOOE*

Class: Frigate
Date: 1742
Location: On and between coral fingers on the ocean side of the reef.

The British frigate HMS *Looe* went down in 1742 under curious circumstances. Some sources have it that she had under tow the disguised French ship, Snow, while other references are unclear on this point. The Looe burned to the water line and her prize, if indeed under capture at the time, joined her on the bottom. Many old commercial sailing vessels and rum runners have since broken their keels on the massive coral ridges. An old anchor from one of the ships runs through one of the coral fingers that has since grown around it. Ballast piles from assorted eras are scattered along the reef, including some 50-pound iron ingots visible in the vicinity of the third coral finger.

NO.15 LOOE KEY NATIONAL MARINE SANCTUARY

Location: From the Bahia Honda Cut, go 6.5 miles (5.6 nautical miles) south-southwest to a 36-foot tower called G Marker. Line up this tower with the 109-foot light tower to the west on American Shoal. Travel 5.1 miles (4.4 nautical miles) west-southwest on the imaginary line until you spot a piling with a triangular sign (Marker 24) mounted on an I beam. Another approach is to take a compass heading of 225-degrees SW of Bahia Honda Channel.

Looe Key is one of the most impressive reefs in the Keys, with huge fingers of living coral that jut out to sea. The area was named after the HMS *Looe*, a ship that ran aground on the shallow reef in 1744 while escorting a captured French ship to South Carolina.

The triangular-shaped shoal is completely awash and varies in depth from 0 to 40 feet. Shallow flats sprinkled with assorted shells lie inshore, and coral fingers hundreds of feet long and up to 15 feet high slope off southward into 40 feet of water. There are many large overhangs on the coral heads, and grouper and barracuda warily swim through the sand-filled valleys. There is also an abundance of macro subjects to keep the underwater photographer busy.

In 1981, a 5-1/2 square mile area surrounding the reef system was designated as the Keys' second marine sanctuary by the National Oceanic and Atmospheric Administration. Active administration of the site began in 1982.

A system of 54 mooring buoys has been placed throughout the Sanctuary to reduce anchor damage and provide a convenient method to secure the many boats that visit the site.

BUOYS NUMBERED 1-32 are located on the central fore reef. These are spectacular coral fingers that have developed over the last 7,000 years. Depths range from one foot at their crest to 30 feet in the sand at their seaward end.

BUOYS NUMBERED 33-39 mark the reef flat on the shore side of the large coral formations. This is an excellent snorkeling spot with depths from 3 to 15 feet. Here, beds of turtle and manatee grass dominate the sea floor. Fascinating coral patches are scattered about. Each harbors a multitude of sea life.

BUOYS NUMBERED 40-49 are located on a system of patch reefs. This site is about one mile north of Marker 24. Depths average 16 feet. Visibility is not as good here as on the outer sections, but the soft coral growth is spectacular.

BUOYS NUMBERED 50-70 are on the intermediate and deep reef sections located southwest of Marker 24. In depths from 30 to 70 feet, enormous sponges and soft corals grow among the low-profile spur-and-groove hard coral formations. At 70 feet, the bottom begins a 30-degree slope seaward to 105 feet. The incline consists of large rocks and ledges inhabited by large fish, and rarer varieties of mollusks, crustaceans and corals. Difficult currents occasionally flow through this area and the depths border on the realm of decompression; explore with care.

The following are regulations and restrictions that are enforced inside the sanctuary. If only a fraction of Looe Key's thousands of visitors fail to observe these rules, the reef's 7,000 year growth will be severely impacted.

- Use the mooring buoys provided, only anchor in sand when buoys are not available.
- Spearfishing is not allowed. This is why you will see so many large fish in the area.
- Hand-feeding of the fish is discouraged. This activity changes the natural behavior of the fish.
- Nothing is to be removed from the sanctuary, including corals, shells, starfish or historical artifacts.
- Regulations prohibiting littering are strictly enforced.
- The diver's red and white down flag must be flown while diving or snorkeling.
- Hook-and-line fishing is allowed. Lobsters may be taken during the season, except in the Core Area marked by the four yellow buoys.
  Fines are imposed for running aground or damaging coral.
  Touching coral with hands, dive equipment or standing is prohibited.

For more information about the Sanctuary, call or write Looe Key National Marine Sanctuary, Route 1 Box 782, Big Pine Key, FL 33043 (305) 872-4039.

Looe Key National Marine Sanctuary Boundary LORAN TD's:

Northwest corner:	**13972.8**	**43538.6**
Southwest corner:	run south from northwest corner to	
	a depth of approximately 145 feet	
Northeast corner:	**13979.4**	**43523.9**
Southeast corner:	run south from northeast corner to	
	a depth of approximately 145 feet	

30

LOWER KEYS

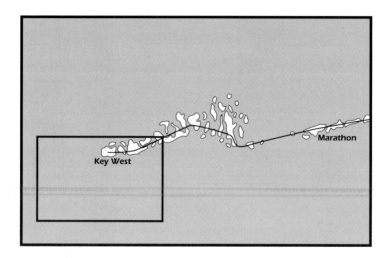

Offshore Diving Locations

1. *USS Wilkes-Barre*
2. Pelican Shoal
3. *The Sambos*
4. *Aquanaut*
5. Ten-Fathom Ledge
6. *All-Alone*
7. #1 Marker Reef
8. Nine Foot Stake
9. Eastern Dry Rocks
10. Eastern Dry Rocks Wreck
11. Rock Key
12. Tile Wreck
13. Sand Key
14. Ten-Fathom Bar
15. Ten-Fathom Ledge off Western Dry Rocks
16. Mallory Square
17. Cottrell Key
18. *Alexander's* Wreck
19. Smith Shoal
20. *Sturtivent* and Other Freighters
21. Marquesas Keys
22. *Northwind*
23. Marquesas Reef Line
24. S-16 WWI Submarine
25. *Cayman Salvager*
26. USS *Curb*

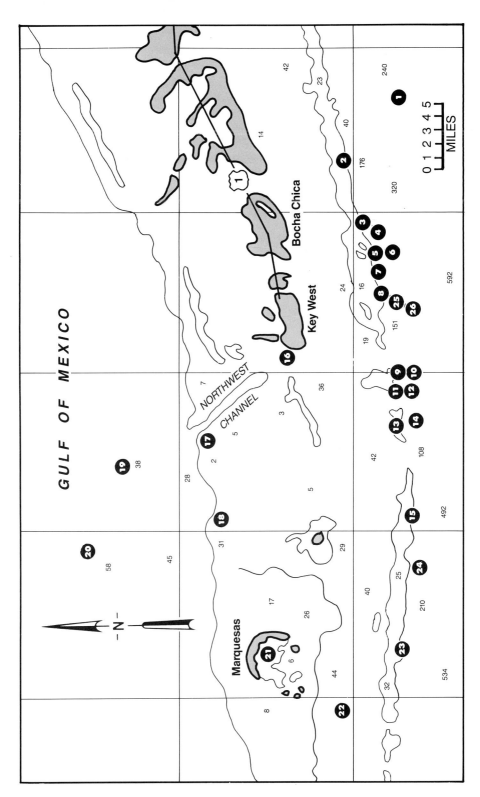

Area Information

For years divers were waylaid by the beautiful reefs of the Upper and Middle Keys and just did not seem to make it to Key West. Today this has changed as divers are discovering the exciting reefs of the Lower Keys. Starting at Western Sambo and continuing 45 miles westward to the Marquesas is a series of spectacular reefs covered with an abundance of sea life. The underwater terrain is dramatic, having towering boulders of coral and sudden drop-offs.

Key West also is a point of departure for the Dry Tortugas, the Marquesas, and other out islands in the Gulf of Mexico, where shell collecting and bug snatching are at their best.

Reef Relief

Key West is the home of Reef Relief, a non-profit group that was organized in 1986 to help protect and preserve the living coral reefs. Their primary goals are: to educate the public not to step, stand or touch the coral; and, the installation of mooring buoys in the Key West waters to reduce anchor damage. If you would like to join this worthwhile organization, send $20 for a yearly membership to: Reef Relief, 1223 Royal Street, Key West, FL 33040.

Mooring Buoys

The following mooring buoys were put in place by Reef Relief in 1988: 25 at Sand Key, 5 at Western Dry Rocks, 10 at Rock Key, 10 at Eastern Dry Rocks, 5 at Western Sambos, and 5 at Pelican. This buoy system will prevent untold anchor damage to the delicate coral.

NO.1 USS *WILKES-BARRE*   13951.8 43570.2
Location: 14 miles west-southwest of the Boca Chica channel.

In 1972, the 610-foot WWII Cleveland Class Light Cruiser *Wilkes-Barre* was sunk to create an artificial reef. The sleek cruiser is completely intact, minus ammo, bunks and fuel. Because of her size, she was placed in 320 feet of water to prevent her from becoming a navigational hazard. She sits on an even keel. The depth to her superstructure is 140 feet; her deck is at 210 feet. Her massive size has done its job of attracting marine life. She teems with schooling fish as well as large bottom-dwellers.

Because of the depths in which she rests, she should be visited only by divers highly experienced in both open-water and deep diving. Local Key West charter boat captains who frequent the site are the only guides who should be considered if you feel yourself capable of making the dive.

NO.2 PELICAN SHOAL
Location: 7 miles west-southwest of the Boca Chica Channel.

This is part of an extensive coral reef line that runs west into the Sambos. Depths range from 10 to 40 feet. Elkhorn and staghorn are the reef's most common corals.

NO.3 THE SAMBOS
Location: 4 miles south of Boca Chica Channel to black and white buoy #2. Looking due east, a pole can be spotted on the east end of Western Sambo.

The Sambos (Eastern, Middle and Western) are divided by expanses of white sand. All sections range in depth from 10 to 40 feet. Good diving is found all along the reef line. A popular spot is in 25 feet of water on the west end of the Western Sambo section. Here, large fields of branch coral grow in profusion. Near this site, at 40 feet, is a series of large, isolated heads covered with sea life. The splendid grouping runs for 125 yards. The grass bed in the shallow water inland from the reefs is good for shelling and lobstering. For best results, maneuver your boat so that it is in 15 feet of water over the grass. Let your boat drift with the slow current and snorkel beside it. Be sure to fly a divers down flag.

NO.4 *AQUANAUT*
Location: 1/2 mile south of Western Sambo.

This is one of Key West's best and easiest wreck dives. A beautiful 50-foot wooden-hulled tug rests in 75 feet of water. She has been down since 1967. When the Gulf Stream sweeps clear water through the area, the tug can be seen in its entirety. What a sight!

NO.5 TEN-FATHOM LEDGE
Location: 1 mile southwest of Western Sambo.

Ten-Fathom is a nice ledge in 35 to 50 feet of water. The series of coral ledges is undercut by several interesting caves noted for hiding big grouper. The ocean side drops from 50 feet to 115 feet. This is a great area for a drift dive.

NO.6 *ALL-ALONE*
Location: On the Ten-Fathom Ledge in 90 feet of water.

The split hull of a 75-foot tugboat rests on the outer ledge. There are always plenty of fish around the old wreck, and grouper and snook are common.

NO.7 #1 MARKER REEF
Location: 4-1/2 miles south of Boca Chica Channel.

This is probably the prettiest shallow reef area off Key West. Fifteen large coral fingers extend out toward the sea.

NO.8 NINE FOOT STAKE
Location: 1 mile west of #1 marker. Marked by a tall pile.

A nice shallow reef in 15 to 30 feet of water. Because the Navy-Air Development Corp. does its testing in this area, divers make all types of interesting finds.

NO.9 EASTERN DRY ROCKS
Location: 2 miles due east of Sand Key and 5 miles south of Key West. Marked in the center by a steel beam.

Large mounds of coral are found southwest of the marker. There is also a wreck on this end of the reef, but it is hard to find due to disintegration of the wood. Divers still find stone ballast and some brass fittings. Lobstering and shelling for conch are good in the shallow water around the marker.

NO.10 EASTERN DRY ROCKS WRECK
Location: Southwestern end of the Eastern Dry Rocks.

Little remains of this old galleon, but an occasional brass fitting is found buried in the white sand.

NO.11 ROCK KEY
Location: 1 mile due east of Sand Key. The area is marked by a steel beam placed in the back center of the reef.

This is a deeper reef with spectacular cracks, some as deep as 20 feet, and only as wide as a diver. There are also good snorkeling and lobstering on the shallow inland side.

Two very old wrecks are located in the area. The first is unidentified. It is located on the southwest end of the reef in approximately 15 feet of water. Among the artifacts recovered are numerous cannon balls and brass spikes.

NO.12 TILE WRECK
Location: Northeast end of Rock Key.

A ship carrying a load of building tiles from Barcelona, Spain went aground on Rock Key. Divers have been digging the tiles out of the sand for years. Other interesting relics which have been recovered include iron ballast, brass spikes and anchor chain.

NO.13 SAND KEY
Location: 6 miles south of Key West, this reef is easy to find. In the middle of the reef there is a small island with a 110-foot lighthouse.

This peaceful sand island is a great place for all-weather diving, as you can always find a lee side with good snorkeling. The best side to dive is the ocean side. When you arrive at the island, go around so that you can see Key West through the lighthouse. Anchor in approximately 15 feet of water and swim toward the island. You will find nice ledges dropping from 45 feet to 70 feet. Numerous grouper are found along the ledges.

NO.14 TEN-FATHOM BAR
Location: 1/2 mile due south of Sand Key. Anchor your boat so that the anchor drops over the wall.

This is a spectacular wall dive with the top of the wall in 25 feet of water and dropping almost straight down to 130 feet. Exceptionally clear water flows through this area much of the year. Divers need to be aware of a strong current which will be encountered the first 30 feet of the dive. If you would like to try your hand at feeding large fish, this is the place to do it. Grouper and snapper are very tame, and it is not uncommon to discover a large grouper looking over your shoulder. Schools of rays have been spotted swimming near the bottom of the ledge. You will see corals and sponges that are not found on the shallower reefs. If you have an underwater tow, be sure to use it here: you will want to explore a great deal more in the limited amount of bottom time you will have.

NO.15 TEN-FATHOM LEDGE OFF WESTERN DRY ROCKS

Location: 4-1/2 miles west-southwest of the Sand Key Light.

This is the hot-spot in Key West waters for reef diving. Most area diving charters spend a great deal of time exploring the coral caves, ledges and outcroppings. Its close proximity to the Gulf Stream increases the water visibility, creating spectacular vistas of underwater coral gardens. The relatively shallow 40 to 50-foot depths add to the popularity of the site. Schools of fish gather around the higher peaks, while a spectacular array of colorful tropicals dance about the undercuts and soft corals. A large black grouper population is another attraction. This is thought to be one of their spawning grounds. Larger grouper are found from March through June.

NO.16 MALLORY SQUARE (Beach Dive)

Location: At the north end of Duval Street in Key West, turn left, go one block, and turn right. The dock on which you find yourself is Mallory Square.

This pier is over 100 years old. It is currently used as a dock for large ships. Diving off the pier in 20 to 30 feet of water has produced thousands of old bottles along with various artifacts dating back to the 1800s. Diving should be done only at slack tide as the currents are very strong during tidal changes. Visibility is in the three- to ten-foot range. An underwater compass is necessary.

NO.17 COTTRELL KEY

Location: Travel out the Northwest Channel for about 9 miles until you come to the ruins of an old Coast Guard lighthouse (only the pilings are standing). Turn west, and run until you are parallel with the first mangrove island on your left. Coast your boat in until you are in 4 to 5 feet of water.

This is a good area for shell collecting. Figs, tuns, milk conch, fighter conch, horse conch, cowries and tulip shells have all been found here. This is also a good area for lobsters. Look for them in the shallow water, hiding in sponges and under rocky ledges.

Cottrell is an old military strafing site. Many shell holes pock the bottom. Artifact hunting is a popular diving activity. Depths range from 10 to 15 feet. This is where divers head when the south wind is too strong for good diving on the Atlantic reefs.

NO.18 *ALEXANDER'S* WRECK   13913.3   43710.8

Location: 4 miles west-northwest of Cottrell Key.

The *Alexander* was a 300-foot Destroyer Escort sunk in the Gulf in 1970. She is in half, lying on her side in 40 feet of water. Visibility at the site averages from 40 to 50 feet. The wreck still produces many military artifacts and is the year round home of swarming baitfish.

NO.19 SMITH SHOAL

Location: 12 miles northwest of Key West. Marked by a 47-foot tower.

Visibility limited to about 20 feet is the only thing that prevents this area from being rated an excellent dive. Huge stands of large plate, brain and staghorn corals loom 15 to 18 feet up from the sand bottom just south of the tower. Depths range from 20 to 40 feet.

NO.20 *STURTIVENT* AND OTHER FREIGHTERS
Location: 7 miles on a 293-degree course from the Smith Shoal Tower.

The broken remains of a four-stack destroyer rest in 70 feet of water. She was sunk in 1942 by our own mines. You can expect visibility of 25 feet. Several other WWII casualties are located in the same general area of the Gulf. They include the 3,018-ton American freighter, *Luchenbach*, that went down with one-sixth of the world's tungston supply and the Norwegian freighter, *Bosikka*, sister ship to the *Benwood*, which is located in the Pennekamp Park nearby. The Danish freighter, *Gumbar*, is considered a hot fishing spot. All of these ships were sunk between June and August of 1942. Fish life is prolific at each, with large grouper, jewfish, cobia and snapper.

NO.21 MARQUESAS KEYS
Location: 25 miles due west of Key West.

The Marquesas are a group of ten mangrove islands surrounded by shallow waters. The area is good for both shelling and spearing. Four miles to the west lies the now-famous Spanish galleon, *Atocha*, successfully salvaged by Treasure Salvors, Inc. of Key West. Ten miles to the west of Halfmoon Shoal is the broken wreckage of an old island ferry, lying in 15 feet of water. These waters are known locally as the Quicksands. Rip currents are constantly present. For this reason, the area is not recommended for novice divers.

NO.22 *NORTHWIND*
Location: 3-1/2 miles southwest of the Marquesas.

A large, metal tugboat belonging to Treasure Salvors tragically sank one night. She rests on her side in 40 feet of water, and is inhabited by several large jewfish.

NO.23 MARQUESAS REEF LINE
Location: A continuation of the Atlantic reef line, running east to west, 6 miles south of the Marquesas.

A long, broken line of coral ledges and caves in 30 to 70 feet of water. Some of the Keys' best spearing is found in these waters. Far from land, seldom dived, large snapper and grouper roam freely here.

NO.24 S-16 WWI SUBMARINE
Location: 14 miles west-southwest of Key West.

The S-16 sub was sunk completely intact in 1945 by the Navy for experimental purposes. She was built in 1919. Her depth of 250 feet makes this a very difficult dive. A few local captains make runs to the area during September or October, when the Keys' weather is at its best.

NO.25 *CAYMAN SALVAGER*   13923.4   43638.7
Location: 1 mile southwest of the Nine Foot Stake.

This 200-foot steel-hulled buoy tender sank in the Key West Navy harbor in the mid-1970s. In 1985, she was being towed out to sea to become an artificial reef when the cable broke. She settled on her side but was righted a few months later by the strong surge created by Hurricane Kate. She now sits on even keel in

90 feet of water less than 20 yards from where the deep reef drops to 130 feet. Her deck is only 60 feet below the surface. The superstructure was removed, but much remains to be explored. Divers are able to penetrate her interior through several openings. Even the engine room is accessible.

NO.26 USS *CURB*   13922.2   43639.7
Location: 1 mile south of the Cayman Salvager.

A naval salvaging tug, 300 feet in length, was sunk November 24, 1983 as an artificial reef site. She sits on even keel in 185 feet. The shallowest point is 120 feet. This is a decompression dive for the experienced only.

THE DRY TORTUGAS

The Dry Tortugas is one of the most varied and beautiful diving locations in Florida. Not only are the diving opportunities numerous and rewarding, but picturesque Fort Jefferson also offers 'landlubbers' a chance for adventure and rare sightseeing.

A brief sketch of the colorful history of Fort Jefferson and the surrounding islands will help familiarize you with the area. The islands were first visited in 1513 by Ponce de Leon, who named the area the "Tortugas" for the great abundance of tortoises he found there. Fort Jefferson stands on Garden Key. Construction was begun on it in 1846 by the U.S. military. The mammoth six-sided, three-tiered fortress was designed for a garrison of 7,500 men and an armament of 450 cannons. Although construction continued for 30 years, she

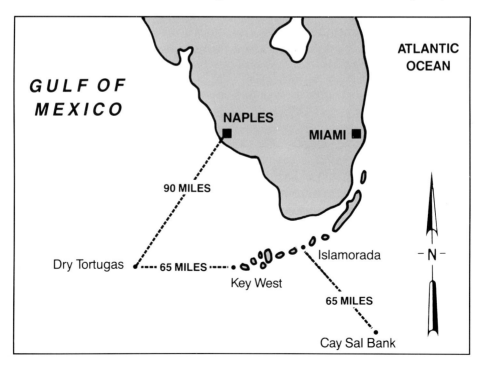

was not completed, and no battle was ever fought from her walls. Following the Civil War, she was turned into a prison. Dr. Samuel Mudd, the physician who set James Wilkes Booth's leg after the assassination of Lincoln, was sent to serve a life sentence at Fort Jefferson after being convicted of conspiracy in the assassination plot. During an epidemic of yellow fever in 1867, the assigned post surgeon was one of the first to die. Dr. Mudd performed admirably by filling in the vacancy, and helped stem the spread of the disease. He was released with full pardon in 1869. His quarters can still be viewed in the fort.

Fort Jefferson covers nearly ten acres. It is approximately half a mile in perimeter, has walls eight feet thick and 45 feet high, and is surrounded by a

Fort Jefferson Monument, Dry Tortugas.   Ned DeLoach

breakwater moat. Fort Jefferson National Monument, along with the surrounding reefs and islands, was set aside on January 4, 1935 as a national preserve by proclamation of President Franklin D. Roosevelt. It is under the jurisdiction of the National Park Service.

As you stand in front of the fort, the island directly east of you is Bush Key, a rookery for thousands of terns, chiefly noddies, sooties and least terns. Anyone wishing to visit the bird sanctuary must acquire permission from the federal park rangers stationed in Fort Jefferson.

Two and one half miles due west of Garden Key is Loggerhead Key, upon which stands the present Coast Guard lighthouse. The lighthouse is 151 feet high. It is the first thing you see as you approach the Dry Tortugas from Key West. The island is almost one mile long and is covered with rugged Australian pines. Just off the beach, on the northwestern side of the island, are beautiful snorkeling areas.

The constant flow of the Gulf Stream current over more than one hundred square miles of living coral reefs guarantees visibility in the 80- to 100-foot range year round. Ledges that drop 30 to 80 feet, 16 reported wrecks, and miles of shallow snorkeling depths invite all types of reef diving enthusiasts. Relative inaccessibility ensures that the Dry Tortugas will seldom be overrun by boaters or divers. The distance from the nearest populated area (Key West) is 65 miles. It can be reached only by boat or seaplane. There are no commercial facilities in the group of eight islands. Most people either sleep in their boats or camp on Garden Key, where there are barbecue grills for cooking. All food and water must be brought in as there is none available. Also, visitors must transport their garbage away with them.

Among the wrecks in the area, one of the favorites is a 300-foot, steel-hulled wreck, located one mile southwest of Loggerhead Key. It is called the French wreck. She is broken in half and lies in 30 feet of water. Many schooling fish make their homes there, among them grunts, snappers, porgies, and thousands of colorful tropical fish. Jewfish may be seen in and under the wreck on almost every dive. Her estimated age is 150 years.

Approximately 100 yards off buoy No. 8 is the wreck of a sailing schooner. She is broken up and very old. This is a good place to look for artifacts. Stone ballast piles lie in 60 feet of water.

A 1971 government survey reported there were 16 wrecks in the area. However, the validity of finds would have to be checked with government officials.

There is an unidentified wreck on White Shoal halfway between Loggerhead Key and Garden Key, lying in approximately 25 feet of water. Several antique bottles have been found which date from the 1840s.

Spearfishing is prohibited within the boundaries of the Monument. You may take fish at depths beyond 70 feet, where you will be out of park limits. Tortugas Bank, seven and a half miles west of Loggerhead Key, is good for spearfishing, with average depths of 50 feet. Two lobsters per day per person may be taken during the season. The taking of other shellfish and coral is prohibited.

If you plan a trip to Dry Tortugas from Key West in your own boat, be sure to check with the National Weather Service in Key West for the latest forecast. Usually discouraging to skippers of small boats is the 30-mile stretch of deep water between Rebecca Shoals and the reef of the Dry Tortugas. Tidal currents, which flow against the prevailing winds, combine with the Gulf Stream current to make this an especially rough area during periods of strong winds, usually in winter. Once reached, however, the harbor in front of Fort Jefferson is well protected.

Dead seashells may be collected when found above the high-water line. Bush Key is reserved for birds during the nesting season of March through September. Then, the sooty terns gather by the thousands from the Caribbean Sea and west-central Atlantic. Their nests are depressions in the warm sand and the parents take turns shading their single eggs from the hot sun. Disturbing the birds during this period can result in the deaths of many of the young. When the babies have finally grown large and strong enough for continuous flight, the entire colony leaves. Brown boobies can also be seen among the sooties, but in much smaller numbers. Frigate birds are sighted in summer, while blue-faced and brown boobies are sighted only occasionally. Songbirds and migrant birds rest at the islands which lie in their flight path from the United States to Cuba and South America. In winter, gulls, terns of the north, and migratory shore birds find refuge here. All keys except Garden and Loggerhead are closed during the turtle season, May through September 30.

Forgotten passage, Fort Jefferson.   Ned DeLoach

An exciting way to visit the Dry Tortugas is by five-passenger sea plane from Stock Island, near Key West. The 35-minute flights leave twice a day. The spectacular ride takes you over sand flats, grass beds, quicksand and steep drop-offs populated by large fish which are easily spotted from the low-flying plane. Schooling tarpon, sharks, rays, turtles and porpoise are sighted on every flight. Directly under the flight path is the hull of the *Northwind*; Treasure Salvors ill-fated tug that helped in the salvage of the treasure-laden Spanish galleon, the *Atocha*. The view of the huge fort, surrounded by crystalline blue waters, is worth the trip by itself. Visitors are given two and a half hours to explore the fort, sunbathe or snorkel over the shallow coral patches west of the fort. The best diving locations are marked by a convenient buoy system.

DIVE BOAT DIRECTORY

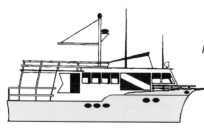

INDEX

Reefs and Ledges

Wrecks, Artificial Reefs and Bridges